95-REM-2

Producing Class A Biosolids With Low-Cost, Low-Technology Treatment Processes

by:

Joseph B. Farrell
Consultant

Douglas T. Merrill
Consultant

Perry L. Schafer
Brown and Caldwell

2004

AF631451

WERF

The Water Environment Research Foundation, a not-for-profit organization, funds and manages water quality research for its subscribers through a diverse public-private partnership between municipal utilities, corporations, academia, industry, and the federal government. WERF subscribers include municipal and regional water and wastewater utilities, industrial corporations, environmental engineering firms, and others that share a commitment to cost-effective water quality solutions. WERF is dedicated to advancing science and technology addressing water quality issues as they impact water resources, the atmosphere, the lands, and quality of life.

For more information, contact:
Water Environment Research Foundation
635 Slaters Lane, Suite 300
Alexandria, VA 22314-1177
Tel: (703) 684-2470
Fax: (703) 299-0742
www.werf.org
werf@werf.org

This report was co-published by the following organizations. For non-subscriber sales information, contact:

Water Environment Federation
601 Wythe Street
Alexandria, VA 22314-1994
Tel: (800) 666-0206
Tel: (703) 684-2452
Fax: (703) 684-2492
www.wef.org
pubs@wef.org

IWA Publishing
Alliance House, 12 Caxton Street
London SW1H 0QS, United Kingdom
Tel: +44 (0) 20 7654 5500
Fax: +44 (0) 20 7654 5555
www.iwapublishing.com
publications@iwap.co.uk

© Copyright 2004 by the Water Environment Research Foundation. All rights reserved. Permission to copy must be obtained from the Water Environment Research Foundation.
Library of Congress Catalog Card Number: 2003113333

IWAP ISBN: 1-84339-693-9 WEF ISBN: 1-57278-184-x

This report was prepared by the organization(s) named below as an account of work sponsored by the Water Environment Research Foundation (WERF). Neither WERF, members of WERF, the organization(s) named below, nor any person acting on their behalf: (a) makes any warranty, express or implied, with respect to the use of any information, apparatus, method, or process disclosed in this report or that such use may not infringe on privately owned rights; or (b) assumes any liabilities with respect to the use of, or for damages resulting from the use of, any information, apparatus, method, or process disclosed in this report.

Brown and Caldwell and its subconsultants

This document was reviewed by a panel of independent experts selected by WERF. Mention of trade names or commercial products does not constitute WERF endorsement or recommendations for use. Similarly, omission of products or trade names indicates nothing concerning WERF's positions regarding product effectiveness or applicability.

ACKNOWLEDGMENTS

In addition to the members of the project team, we would like to acknowledge the contributions of many people, especially: Metropolitan Water Reclamation District of Greater Chicago (Cecil Lue-Hing, Prakasam Tata, Sam Sedita, Jim Zmuda); Sacramento County, California (Stan Dean, Ruben Robles, Brian True); City of San Jose, California (Ron Garner, Steve Schrag); and Miami-Dade Water and Sewer District (John Chorlog, John Kabouris).

Report Preparation

Principal Investigator:
Perry L. Schafer, P.E., DEE
Brown and Caldwell

Project Team:
Joseph B. Farrell, Ph.D., P.E.
Consultant

Douglas T. Merrill, Ph.D., P.E.
Consultant

Project Subcommittee

David DiGregorio, P.E., Chair, *Eimco Process Equipment Company*
Joel E. Alpert, Ph.D., *Consultant*
Steven K. Dentel, Ph.D., P.E., DEE, *University of Delaware*
Cecil Lue-Hing, D.Sc., P.E., DEE, *Cecil Lue-Hing Associates*
James E. Smith, Jr., D.Sc., *U.S. Environmental Protection Agency*
Prakasam Tata, Ph.D., QEP, *Tata Associates International*
Salvador J. Sedita, Ph.D., *Consultant*
James T. Zmuda, Ph.D., *Metropolitan Water Reclamation District of Greater Chicago*

Water Environment Research Foundation Staff

Director of Research: Daniel M. Woltering, Ph.D.
Project Manager: Bonnie Bailey
Christy Bixler, P.E. (as of September 2003)

ABSTRACT AND BENEFITS

Abstract:

Class A biosolids can be produced using low-cost, low-technology biosolids treatment processes including lagoon storage, air drying, and cake storage. This project reviewed the available literature and municipal agency data about these processes. This report presents design and operating guidelines distilled from the review process. It is designed for wastewater treatment plant (WWTP) managers, operators, and engineers who wish to discern whether these processes, used alone or in combination, might be practically applied at specific plants. This report also describes the U.S. regulatory environment in relation to producing Class A biosolids and defining Class A processes. It also presents a list of recommended research needs.

Benefits:

- Familiarizes WWTP managers, operators, and engineers with low-cost, low-technology biosolids treatment processes, likely pathogen kill mechanisms, and practices that have reduced pathogen densities to Class A levels at scales ranging from laboratory tests to large municipal biosolids treatment operations.

- Presents guidelines for producing Class A biosolids under a variety of conditions.

- Describes low-technology treatment processes within the Class A regulatory framework, identifies satisfactory end conditions for products created from low-tech treatment processes, and provides guidance in developing national or site-specific certification as processes equivalent to a process to further reduce pathogens (PFRP).

Keywords: Pathogens, Class A, helminth ova, virus, salmonellae, fecal coliform, lagoons, air drying, cake storage, Part 503, PFRP equivalency, biosolids

TABLE OF CONTENTS

LIST OF ACRONYMS

CFR	Code of Federal Regulations
DS	dry solids
g	grams
HBBMP	Hornsby Bend Biosolids Management Plant
HSSPT	high-solids sludge processing train
LSSPT	low-solids sludge processing train
log(s)	logarithmic base 10, i.e., $\log_{10}$
MPN	most probable number
MWRD	Metropolitan Water Reclamation District of Greater Chicago
PEC	Pathogen Equivalency Committee
PFU	plaque-forming units
PSRP	process to significantly reduce pathogens
PFRP	process to further reduce pathogens
SOUR	specific oxygen uptake rate
SPT	solids processing trains
SRT	solids retention time
TSS	total suspended solids
U.S. EPA	United States Environmental Protection Agency
VAR	vector attraction reduction
VSS	volatile suspended solids
WWTP	wastewater treatment plant

Executive Summary

In 1984, the United States Environmental Protection Agency (U.S. EPA) established a policy to promote the beneficial use of treated municipal sludge (called "biosolids" hereafter). Today, biosolids are viewed as valuable manufactured products, as opposed to waste. The government has established two biosolids classifications: A and B.

Standards for Class B biosolids are less stringent than for Class A. Class B biosolids must contain fecal coliform densities of <2 million most probable number (MPN)/g dry solids and meet vector attraction reduction (VAR) requirements. Class B biosolids are subject to site restrictions that limit crop harvesting, animal grazing, and public access.

Class A biosolids must also satisfy VAR requirements while achieving higher standards for pathogens and pathogen indicators, specifically:

- Fecal coliform limits of <1000 MPN/g dry solids or salmonellae limits of <3 MPN/4 g dry solids.
- Enteric virus limits of <1 plaque-forming unit (PFU)/4g dry solids.
- Helminth limits of <1 viable ovum/4 g dry solids.

Class A biosolids may be sold or given away without any pathogen-related restrictions because their pathogen densities are very low (at detection limits). Their high quality enhances their usefulness, value, and general applicability.

Class A biosolids can be produced by sludge treatment trains employing one or more low-technology, low-cost treatment process (lagoon storage, air drying, and cake storage). While these processes are commonly perceived to be most applicable at small wastewater treatment plants (WWTP), they are also used by large facilities in major cities to produce Class A biosolids (Chicago, Illinois, and San Jose, California, for example).

ES.1 Objectives of this Report

The objectives of this report are to:

- Familiarize operators of WWTPs with the lagoon storage, air drying, and cake storage processes.
- Summarize the available literature regarding these processes, note what has worked well and what has not, and define trends, if possible.
- Describe possible operating conditions necessary to produce Class A biosolids through lagoon storage, air drying, and cake storage.
- Describe the U.S. regulatory environment in which these processes must operate.
- Define further research needs.

ES.2 Process Descriptions

Low-technology sludge treatment processes and likely pathogen kill mechanisms are described in Sections ES.2.1 through ES.2.3.

ES.2.1Storage Lagoons

Storage lagoons receive biosolids (liquid and/or dewatered digested sludge) either continuously or intermittently. The solid materials settle to the lagoon bottom, and supernatant is most often returned to the WWTP. After the lagoons fill with biosolids, they are left unfed for a period of time, allowing for further biosolids decomposition and pathogen destruction. Harvesting of the stored biosolids (usually by dredge) can take place after the pathogen content has dropped to desired levels. Mechanisms that may be used destroy pathogens of concern in lagoon storage systems are a continuation of catabolic activity of the digestion process, thermal effects, suppression by nonpathogenic bacteria (which compete with pathogenic organisms for available food), and chemical action (as might be exerted by ammonia and/or organic acids).

ES.2.2Air Drying

In impervious-pad air-drying operations, digested, dewatered biosolids are typically applied to the pad. When the biosolids have dried sufficiently (biosolids content reaches 30% to 35%), piles can be shaped and mounded in windrows. Windrows are usually about 2 ft high, although they can be higher. Water evaporates from the windrows. Periodic turning by mechanical equipment accelerates drying by introducing dry air into the windrow, releasing moist air, and replacing dry particles on or near the surface with moist particles from the interior. Turning also provides oxygen to oxidize remaining biodegradable material, sometimes generating enough heat to elevate windrow temperature. Higher than ambient temperatures speed the process and accelerate pathogen kill. Desiccation may also play a part in pathogen kill. The presence of nonpathogenic bacteria and chemical action can further suppress pathogen activity.

Less frequently, liquid biosolids (slurries) are applied to the pad to depths of approximately 1 ft to 3 ft and supernatant is drained off or allowed to evaporate. When the biosolids have drained or dried sufficiently, they can be shaped and mounded in windrows and drying can proceed, accelerated by mechanical turning as described above.

Alternatively, liquid biosolids can be dried on porous media, typically sand beds. Liquid biosolids are applied to the beds in depths up to 1 ft. The beds drain and dry over a period of one to several months, depending on ambient conditions. Typically, biosolids are left on sand beds until their solids content is about 30% to 50%. Then the material can be readily removed. In humid climates, further drying is likely to be slow. If a Class A product is desired, biosolids may need to be removed from the bed and further dried in windrows. In dry, hot climates, however, the biosolids can remain on the beds, shrinking to a thin layer of very high solids content. High pathogen kill levels can be reached by heating this thin layer completely through to high temperatures on a sunny day, particularly if the biosolids rest on a dark-colored surface.

ES.2.3Cake Storage

Cake storage is similar to air drying on impervious pads, except the windrows tend to be larger and windrow turning is minimal. The cake is stored until pathogen densities have dropped to Class A levels. There is no intent to produce a dried product, although some drying may occur. Pathogen destruction mechanisms are the same as for windrow drying, except desiccation plays a minor role.

ES.3 PFRP Equivalency

Criteria for fecal coliform **or** salmonellae must be satisfied at **all** times, regardless of the process used. However, certain Class A alternatives (Alternatives 1, 2, 5, and 6; see Appendix F) are assumed to satisfy virus and helminth ova criteria by achieving specified process conditions. Operators using these processes need not monitor viruses and helminth ova because experience has shown Class A criteria for these pathogens are satisfied if the specified process to further reduce pathogens (PFRP) conditions are met. For example, thermal processes (Alternative 1) must achieve certain time–temperature requirements, and alkaline treatment processes (Alternative 2) must satisfy certain time–pH–temperature conditions.

Facilities that use lagoon storage, air drying, and cake storage processes are generally not exempt from virus and helminth monitoring. These processes normally fall within Alternative 4 (unknown processes) because the conditions that can achieve successful, reproducible pathogen kill have not been adequately defined. Therefore, facilities using lagoon storage, air drying, and cake storage as part of a treatment train to produce Class A biosolids must monitor viruses and helminth ova as well as fecal coliform or salmonellae in every pile or batch of biosolids produced.

There has been interest on the part of some operators of lagoon storage, air drying, and cake storage processes to achieve process approval within Alternative 6, processes equivalent to a PFRP. The prime sources of this interest are having a defined and approved process that produces Class A biosolids, reduces pathogen testing costs, and increases public confidence in biosolids. Helminth ova and virus testing are expensive and require specialized skills. With PFRP equivalency, these analyses would not be needed. Besides reducing analytical costs, receipt of an official declaration of equivalency substantially enhances public acceptance and willingness to support use of the treated product.

The regulations identify the U.S. EPA regional staff as the authorities who decide whether a process has been demonstrated to be equivalent to a PFRP. U.S. EPA has established a committee of experts, the Pathogen Equivalency Committee (PEC), which provides recommendations to the regulatory staff.

The PEC has provided guidance on equivalency. The PEC requests that applicants for equivalency show that their sludge-treatment process train produces a reduction in enterovirus density of three or more logs (that is, a reduction by a factor of 1000 or more) and a reduction in viable helminth ova density of two or more logs (that is, a reduction by a factor of 100 or more). The PEC has provided recommendations of PFRP-equivalency that fall into two categories: national equivalency and site-specific equivalency.

A national equivalency process can be used anywhere in the country without monitoring of virus and helminth ova. However, qualifications may be included in the approval. For example, for one process that is conducted outdoors, the ambient temperature must be above a specified value. This process can be used nationally wherever the temperature requirement is met.

Site-specific recommendations are given when U.S. EPA has reservations that the process may not be applicable to the range of feedstocks or other conditions that could be encountered in the United States. Then, U.S. EPA might restrict its approval to the site that is applying for certification. The agency has also provided equivalency that is not only site specific, but requires some monitoring of either viable helminth ova or enterovirus densities of the final product. This limited approval might be granted when the U.S. EPA is not confident that the organism in question is not consistently destroyed by the process.

The application process for PFRP equivalency can be difficult, lengthy, and expensive. Agencies that apply for certification can speed the process by providing U.S. EPA with a complete description of its process(es) and the effects of treatment parameters on process performance. Early and continuous communication must be maintained with U.S. EPA regarding the agency's expectations for sampling, quality assurance/quality control requirements, and the range of each parameter that should be investigated. The level of confidence in the results that U.S. EPA requires may substantially exceed what the proposing organization expects. Initial communication with U.S. EPA is necessary to obtain an appreciation of the magnitude and standards of the proof required by the agency.

National certification under PFRP equivalency can be viewed as a destination for low-tech processes, where process mechanisms are well enough understood that certain operating conditions can satisfy Class A requirements and virus and helminth monitoring are not needed. This document provides guidance toward that goal.

ES.4 Treatment Guidelines

Distillation of the information in this project's literature review (presented in Appendices A through E) yields *approximate* operating guidelines for low-technology sludge treatment processes. The guidelines should be used with discretion and caution. They are empirical, evolving from observations made under a variety of conditions including laboratory-, pilot- and full-scale operations. Results obtained at one location may not transfer well to another. This is because the mechanisms responsible for pathogen kill in these low-technology processes are not completely understood and, therefore, are difficult to quantify; feedstock quality varies between locations; and plant operators may not use identical operating procedures. Consequently, the reader should not assume that using these guidelines ensures production of Class A product. Pathogen monitoring is required. Nevertheless, the guidelines should help those considering using these processes make reasonable design and operating decisions.

The guidelines describe conditions believed to be necessary for achieving the Class A criterion for viable helminth ova for processes operated at or near ambient temperatures. Helminth ova are reportedly the hardiest of the regulated pathogens found in municipal sludge. Reduction of viable helminth ova to levels stipulated in the regulations most likely ensures destruction of the other regulated pathogens. The destruction of helminth ova thus typically dictates processing times for lagoon storage, air drying, and cake storage.

All guidelines assume the low-technology processes are preceded by effective biological digestion. The literature suggests that anaerobic digestion is more effective than aerobic digestion in this pretreatment role; that is, low-technology processes inactivate pathogens better when combined with anaerobic digestion than they do when combined with aerobic digestion. The higher concentrations of ammonia and organic acids in anaerobically digested biosolids, which carry through the low-tech processes, may partially account for this result.

ES.4.1 Guidelines for Lagoon Storage

Laboratory and pilot-scale lagoon storage systems operating at 25°C (77°F) have reduced the density of viable *Ascaris* ova in *Ascaris*-spiked, anaerobically digested biosolids to below the

Class A criterion within 5 to 16 months. The densities of viable *Ascaris* ova were reduced significantly (3.7 to 4.4 log reduction) during this process. Therefore lagoons operating at 25°C (77°F) and above appear capable of producing Class A biosolids within 16 months under most conditions. The human pathogen *Ascaris lumbricoides* is screened for when assessing the presence of viable helminth ova in biosolids. *Ascaris* was chosen as the parasite indicator organism in the 1970s when the U.S. EPA regulations were being written because these helminth ova have a long survivability in the environment and are easy to identify due to size.

Laboratory storage of spiked, digested biosolids at 4°C (39°F) for 25 months reduced the density of viable *Ascaris* ova by less (0.8 to1.6 log reduction). This was insufficient to achieve the Class A criterion. However, if the log reduction performance demonstrated in these laboratory scale experiments can be maintained with the much lower *Ascaris* densities found in typical municipal biosolids (and there is no certainty that it can be), then lagoons operating at 4°C (39°F) might produce Class A biosolids within 25 months, provided the densities of viable helminth ova in lagoon feed are relatively low (≤10 ova/4g dry solids).

There is less information about the rates of inactivation of *Ascaris* ova in the temperature range between 4°C and 25°C (39°F and 77°F) and therefore definitive storage-time guidance cannot be given for this temperature range. Such information is needed, since most storage lagoons operate in that range.

ES.4.2 Guidelines for Air Drying and Cake Storage

Desiccation, temperature, and time are the primary pathogen-kill mechanisms for air drying systems.

ES.4.2.1 Desiccation-oriented Drying Systems

It is necessary to reduce biosolids moisture contents to ≤5% to achieve complete inactivation of helminth ova by desiccation. Operators of most U.S. air drying plants will be unable to dry their biosolids to these moisture levels because of rainfall or humidity. Operators in hot, arid regions like the southwestern U.S. may be the exception. However, even those who can achieve such low moisture levels may be unwilling to do so because the final product may be too dusty. Drying time and drying temperature still play roles in desiccation-oriented drying systems. However, no time and temperature guidance can be given for these systems at this time.

ES.4.2.2 Time- and Temperature-dependent Systems

Air drying systems that do not rely on desiccation (the great majority) and cake storage systems must provide processing times long enough and operating temperatures high enough to ensure the requisite destruction of helminth ova. These air drying or cake storage systems should be operated for at least 250 days in locations with warm summers and mild winters and at least 350 days in locations with warm summers and cold winters. A summer should be included in the operating period. Pile or windrow moisture content should remain above approximately 40% for most of the storage or drying periods to promote continued biological activity. Windrows should be turned to control the windrow moisture content and temperature.

ES.4.3 Guidelines for Coupled Lagoon Storage/Air Drying Systems

Lagoons can be used in combination with air drying to produce Class A biosolids. Because organisms exit the lagoons in a debilitated state, they are vulnerable to the altogether different environment they encounter in the air drying process. This sequential combination is probably more powerful than lagoon storage or air drying processes operated alone when temperature conditions are the same for all. Therefore, guidelines for sequential lagoon storage and air drying systems are different from guidelines presented for those systems when operated individually.

The Metropolitan Water Reclamation District (MWRD) of Greater Chicago treatment train serves as a model for combined lagoon storage and air drying systems. MWRD's lagoons operate with a minimum of 18 months of unfed storage time. The air drying systems that follow are operated during the warm-weather drying season to produce a 60% solids product within 9 to 10 weeks. Windrows are turned three times a week. Treatment times for other coupled systems may be less in warmer climates, more in colder climates.

ES.5 Research Needs

Following is a list of research needs:

- Develop practical methods for assessing helminth kills in full-scale lagoon storage, air drying, and cake storage systems and the time/temperature conditions necessary to achieve these kills.
- Determine temperature/moisture profiles in biosolids windrows, drying beds, and cake piles.
- Develop conclusive data that viruses in municipal sludge die off faster than viable helminth ova.
- Develop a consensus for an improved and more-accepted assay for quantifying viable and nonviable helminth ova in biosolids.
- Evaluate different approaches to increase the concentration of un-ionized ammonia in systems designed to produce Class A biosolids, including adjusting pH and possibly adding urea or other compounds.
- Develop a practical method for determining the potential for regrowth of fecal coliform or salmonellae in stabilized biosolids product.

CHAPTER 1.0

INTRODUCTION

Lagoon storage, air drying, and cake storage are low-cost, low-technology sludge[1] treatment methods. When combined with other pathogen-reduction processes, such as digestion, use of these technologies can reduce pathogen densities so that the treated product (known as biosolids) meets the Class A requirements of Subpart D of 40 Code of Federal Regulations (CFR) Part 503 regulations. When these requirements are satisfied, the treated biosolids can be applied on land without any pathogen-related restrictions.

1.1 Objectives of this Report

The objectives of this report are to:

- Familiarize operators of wastewater treatment plants (WWTPs) with the lagoon storage, air drying, and cake storage processes.
- Summarize the available literature regarding these processes, note what has worked well and what has not, and define trends, if possible.
- Describe possible operating conditions that will produce Class A biosolids through lagoon storage, air drying, and cake storage.
- Describe the regulatory environment in which these processes must operate.
- Define further research needs.

1.2 Organization of this Report

This report is divided into an executive summary, six chapters, and seven appendices.

- Chapter 1.0 describes the report's scope, objectives, and organization.
- Chapter 2.0 describes the lagoon storage, air drying, and cake storage processes and their likely mechanisms of pathogen kill.
- Chapter 3.0 discusses regulations pertaining to production of Class A biosolids using these processes.

[1] Sewage sludge is the residue generated during treatment of domestic sewage. Biosolids refers to sewage sludge that has undergone treatment and meets federal and state standards for beneficial use (Class A or Class B biosolids).

- Chapter 4.0 presents U.S. EPA Pathogen Equivalency Committee (PEC) guidelines for WWTP operators who wish to have their biosolids processing trains certified as equivalent to a process to further reduce pathogens (PFRP).
- Chapter 5.0 suggests operating guidelines for production of Class A biosolids by lagoon storage, air drying and cake storage, and combined lagoon storage/air drying systems.
- Chapter 6.0 discusses additional research needs.

The report also contains seven appendices:

- Appendix A reviews the lagoon storage literature.
- Appendix B reviews the air drying literature.
- Appendix C reviews the cake storage literature.
- Appendix D reviews the literature on combined lagoon storage/air drying systems.
- Appendix E reviews the literature on the effect of ammonia on pathogen kill.
- Appendix F provides an overview of Subpart D of the Part 503 regulation.
- Appendix G explains how the operating guidelines were developed in Chapter 5.0.

CHAPTER 2.0

PROCESS DESCRIPTIONS

This chapter describes the lagoon storage, air drying, and cake storage processes and their variations. It also discusses mechanisms by which they promote pathogen kill.

2.1 Lagoon Storage

Storage lagoons hold anaerobically or aerobically digested biosolids over long periods, typically 1 to 4 years. During storage, biosolids decompose anaerobically and pathogens are destroyed.

2.1.1 Storage Alternatives

Typically, digested biosolids are fed either continuously or intermittently to the lagoon. Biosolids settle to the bottom if the lagoon is operated with a liquid cap layer. Supernatant often is returned to the treatment plant. Facultative lagoons operate with anaerobic bottom layers and aerobic upper layers created by surface aeration. Anaerobic lagoons are anaerobic throughout.

After they are filled, the lagoons are allowed to sit unfed for a period of time, and further solids decomposition and pathogen destruction occur. Harvesting of the stored biosolids (by dredge or other means) can take place after the pathogen content has dropped to levels specified in the Part 503 regulations.

However, some lagoon storage systems use somewhat different operating protocols. For example, the Metropolitan Water Reclamation District (MWRD) of Greater Chicago operates high-solids sludge processing trains (HSSPT) and low-solids sludge processing trains (LSSPT). Storage lagoons are used in each train (Tata et al., 2000).

The HSSPT starts with mesophilic anaerobic digestion. Digester product is treated with a cationic polymer, then centrifuged to produce a 25% solids cake. The cake is transported by rail and placed in HSSPT lagoons, except when temperatures fall to $\leq$-3.9°C (25°F). At these low temperatures, the cake freezes in the rail cars. During such low-temperature days, the centrifuges are not operated and digester product is pumped directly to LSSPT lagoons. The centrifuged cake is stored and aged in the HSSPT lagoons to further stabilize the solids and inactivate pathogenic organisms. After storage for a certain amount of time, the cake is removed from the lagoons and air dried to 60% solids.

The LSSPT lagoons are filled in cycles over one to three years. Each addition of biosolids is allowed to thicken by settling and evaporation before the next increment is added. Supernatant is returned to the wastewater reclamation plant. The layering process continues until the lagoons are filled. The biosolids are then further dewatered using the slackline process (U.S. EPA, 1979). Next, the material is aged in the lagoons to achieve further solids stabilization and disinfection. Finally the aged biosolids are removed from the lagoon and air dried to a solids concentration of 60%.

2.1.2 Potential Pathogen-reduction Mechanisms with Lagoon Storage

Mechanisms that might be expected to destroy pathogens of concern in lagoon storage systems are chemical action, thermal effects, storage time, and suppression by non-pathogenic bacteria.

The un-ionized form of ammonia (NH_3) has been shown to be a potent virucide and ovacide (see Appendix E). Anaerobically digested biosolids contain an abundance of ammonia, but at the pH of typical drying operations (5.0 to 7.0) the un-ionized fraction is very small compared to that of the non-lethal ionized form (NH_4^+). Nevertheless, this small fraction may play a key role in inactivation. The un-ionized fraction can be increased by increasing the pH. For example, the un-ionized fraction is about 0.006 at pH 7; at pH 9 it is about 0.36. The greater concentration of un-ionized ammonia at higher pH is likely to accelerate pathogen kill. Aerobically digested biosolids contain little ammonia since most is normally nitrified.

Elevated or even ambient temperature and time of exposure are known to reduce the densities of all pathogens. There is considerable information about the time-temperature relationship for all pathogens of concern in sludge (see Chapter 5.0 and Appendices A, B, C, and D).

Non-pathogenic bacteria repress the growth of pathogenic bacteria by competing for available food and possibly through enzymatic action. An ample supply of non-pathogenic bacteria exists in the sludge streams of most biological WWTPs, providing the competition needed in downstream lagoon storage, air drying, or cake storage systems.

2.2 Air Drying

Air drying typically produces materials with final solids concentrations of about 50% to 70% (30% to 50% moisture). Drier products can be obtained in hot and arid climates but are often avoided because of their dusting potential. Air drying can be conducted with biosolids piled up in windrows on impervious pads or with biosolids in a single uniform layer on a porous bed, usually sand.

2.2.1 Drying in Windrows on Impervious Pads

Typically, dewatered biosolids are applied to the pad. When the biosolids have dried sufficiently that piles can be shaped (piles can be shaped when the solids content reaches about 30% to 35%), they are mounded in windrows. Windrows are usually about 2 ft high, although they can be higher. A windrow's cross-section approximates an isosceles triangle, with the sides at a slope of about 30° from the horizontal. Initially air penetrates the windrow only a few inches. Air penetrates deeper as the windrow becomes drier. Periodic turning by mechanical equipment accelerates the drying process. Turning introduces dry air into the windrow, releases moist air, and replaces dry particles on or near the surface with moist particles from the interior.

Even if the biosolids have undergone long-term lagooning, considerable degradable organic material remains. This organic matter can be biologically oxidized when exposed to air, generating heat. The internally generated heat can cause the windrow temperature to rise, which not only accelerates the dewatering process but also helps to destroy pathogens. As the pile dries, biological oxidation slows down, but desiccation may continue to kill the pathogens.

Some operators have blended a portion of the dried product with freshly dewatered biosolids to produce an initial pile solids content of 30% to 40%. This action eliminates the period of initial drying needed to develop piles that can be shaped. It can also result in immediate heat generation and reduce odor potential.

As noted previously, impervious pads typically receive dewatered biosolids. Less frequently, liquid biosolids (slurries) are applied to depths of approximately 1 ft to 3 ft and supernatant is drained off or allowed to evaporate. When the biosolids have drained or dried sufficiently that they can be shaped, they are mounded in windrows and drying proceeds with mechanical turning as previously described.

Drying biosolids in windrows has a superficial resemblance to windrow composting. Experience with windrow composting shows that high temperatures can be generated internally which in turn can inactivate pathogens. However, windrow composting typically takes place over a relatively short period (approximately three weeks) with the moisture content ranging from 40% to 60%, the use of bulking agents like wood chips to improve windrow porosity, and frequent turning of the windrows to provide oxygen for the reactions that generate heat.

Air drying of biosolids can generate sufficient heat to initiate smoldering and combustion. Factors that affect this include pile moisture conditions, climate conditions, type/degree of prior sludge/biosolids processing, degradable organic and chemical content, and pile size (larger piles allow more heat buildup). Monitoring interior pile temperatures and controlling critical conditions may be necessary to avoid this problem. Heat generation and temperature rise within drying piles is helpful for pathogen destruction, but too much heat buildup is a potential fire and safety concern.

2.2.2 Drying on Porous Media

Liquid biosolids (slurries) are applied to porous media beds in depths up to 1 ft, and then the bed drains and dries over a period of one to several months, depending on ambient conditions. Typically, biosolids are left on sand beds until their solids content is above approximately 35% and they can be removed by shovels or plows as a solid material that can hold its shape. Under these conditions, the bulk of the material will not get hot enough or dry enough for the biosolids to meet Class A pathogen reduction requirements. In humid climates, further drying is likely to be slow (east of the Mississippi River, for example). To improve chances of achieving Class A materials, biosolids normally must be removed from the bed and further dried in windrows, as described in the previous section on drying on impervious pads.

In dry, hot climates, however, the biosolids can be left on the beds and thus can reach very high solids contents. The biosolids shrink in mass in proportion to the amount of water evaporated. For example, they shrink by a factor of 17 when they dry from 4% solids to 68% solids. Thus, a 12" layer of biosolids could shrink to <1"in thickness. A layer this thin can be heated completely through to high temperatures on a sunny day, particularly if it rests on a dark-colored surface.

The nature of the escape of water during the drying operation also could lead to high temperatures in the solids layer. The entire surface of the bed can receive heat from the sun and the air. Generally, the surface seals over and most of the moisture escapes through cracks in this sealed layer. This unbalance (large area for heat transfer in, small area for diffusion of moisture out) allows for development of high temperatures in the solids layer.

2.2.3 Potential Pathogen Reduction Mechanisms with Air Drying

Mechanisms that might be expected to destroy pathogens of concern are desiccation, chemical action, thermal effects, drying time, and suppression by non-pathogenic bacteria.

When moist biosolids are desiccated (substantially dried), any salts dissolved in the aqueous phase are concentrated. This concentration generates an osmotic effect that collapses and destroys vegetative bacterial cells by drawing water out of them. This osmotic effect, as well as the effect of the low vapor pressure of water during desiccation, can remove water of hydration from the chemicals that make up viruses, spores, cysts, and helminth ova and could possibly destroy them.

The effects of chemical action, thermal conditions, drying time, and suppression by non-pathogenic bacteria were discussed in Section 2.1.2.

2.3 Cake Storage

Dewatered cake can be stored in high windrows or piles until pathogen and fecal coliform densities decrease to levels required to satisfy Class A requirements. There is no intent to produce dried biosolids by this method, although some drying may occur. Piles are normally allowed to stand without being turned.

For purposes of discussion, a high windrow is arbitrarily defined as one over 4ft in height. It is possible to store cake in windrows if the solids content is over about 30% to 35% solids. If the solids content is much less than this, a trench or retaining walls are likely to be needed to contain the cake.

The transport of water and air in such windrows is by diffusion (which is very slow), and migration of bacteria is hindered. The windrows will lose moisture to the atmosphere from its outer surface unless they are covered. Conditions will be non-uniform, making it difficult to define a process, which is a necessary requirement for achieving designation as a process equivalent to a PFRP (see Chapter 4.0 for information about processes equivalent to PFRPs).

Covering a windrow such as this with polyethylene sheeting has some merit. Evaporation will be reduced so the windrow will remain at uniform moisture content and the windrow will not be cooled by evaporation or its exterior washed away or weathered by the environment. With covering, the temperature may rise somewhat above the temperature of the surroundings, increasing the likelihood of pathogen destruction. Naturally, little drying will take place, so either the material will have to be used in this condition or dried further in another operation. However, there is some possibility that a covered windrow or pile might get hot enough as to smolder and eventually catch fire.

If the windrow is left unturned by mechanical means, the process will not be uniform since the interior material is clearly treated differently than the material in the outer shell of the windrow. Turning helps make treatment and the final product uniform. Periodic turning provides oxygen for oxidation of organic matter, which in turn can generate heat and raise the pile temperature. Turning also exposes pathogens on the windrow surface to the heat and moist conditions of the windrow interior and pathogens in the interior to the desiccation of the exterior surface.

Pathogen kill mechanisms are likely the same for cake storage as for air drying, except desiccation plays a relatively minor role.

2.4 Bacterial Regrowth

Fecal coliforms and salmonellae remaining after a treatment process or inoculated from outside sources (such as birds, insects, and other batches of contaminated solids) can grow again if conditions to support regrowth are present. These conditions include:

- Availability of food

- Presence of sufficient moisture
- Insufficient inhibitory chemicals
- Insufficient non-pathogenic organisms—such organisms compete for food with pathogenic bacteria and destroy pathogens by enzymatic action.

If these conditions occur, fecal coliform and salmonellae populations initially satisfactory as the result of treatment can regrow to levels exceeding Class A requirements. Therefore, biosolids treatment and storage arrangements should be set up so that one or more of the above conditions are not present. Chapter 5.0 describes methods for preventing bacterial regrowth in lagoon storage, air drying, and cake storage systems.

Viruses and helminth ova do not regrow. Their densities can never increase above the level provided by treatment.

CHAPTER 3.0

Regulations that Apply to Lagoon Storage, Air Drying, and Cake Storage

All systems that attempt to produce Class A or Class B biosolids are regulated by Subpart D of the 40 CFR Part 503. This chapter provides an overview of the regulation as related to production of Class A biosolids by lagoon storage, air drying, cake storage, and combinations thereof. Appendix F provides more detail.

3.1 Class A Criteria for Pathogen Reduction

If product from *any* treatment train (including trains that contain lagoon storage, air drying, or cake storage processes) is to achieve Class A status, it must satisfy the following criteria:

- Either the density of fecal coliforms must be <1000 most probable number (MPN)/g of total solids (dry weight basis) *or* the density of salmonellae bacteria must be <3 MPN/4 g of total solids (dry weight basis) at the time the biosolids are used or disposed; at the time the biosolids are prepared for sale or given away in a bag or other container for application to land; or at the time the biosolids or material derived from biosolids are prepared to meet the requirements in Part 503.10(b), (c), (e), or (f).
- The density of enteric viruses must be <1 plaque-forming unit (PFU)/4 g of total solids (dry weight basis).
- The density of viable helminth ova must be <1 viable ovum/4 g of total solids (dry weight basis).

The criteria for fecal coliform or salmonellae must be satisfied at *all* times, no matter the process used.[1] However, certain Class A alternatives (Alternatives 1, 2, 5, and 6; see Appendix F for details) are assumed to satisfy virus and helminth ova criteria by achieving specified process conditions. Operators using these processes need not monitor viruses and helminth ova because experience has shown the Class A criteria for these pathogens are satisfied if the specified process conditions are met. For example, thermal processes (Class A Alternative 1) must achieve certain time and temperature requirements and alkaline treatment processes (Alternative 2) must satisfy certain time-pH-temperature conditions. Exemption from virus- and helminth-monitoring requirements is highly desired, because virus and helminth ova analyses are time-consuming and expensive.

[1] It is not necessary to satisfy both fecal coliform and salmonellae criteria. Meeting the criteria of one is sufficient to satisfy Part 503 rules.

Facilities that use lagoon storage, air drying, and cake storage processes are generally not exempt from virus and helminth monitoring. These processes fall within Class A Alternative 4 processes (unknown processes) because the conditions that can achieve successful, reproducible pathogen kill in these processes are not adequately defined. Reasons for this lack of definition include:

- Uncertainty about the relative importance of the various pathogen kill mechanisms that are critical for these processes. This uncertainty leaves process designers, regulators, and operators unsure of the process parameters most important to control.
- Inability to control environmental conditions that affect some of these mechanisms (e.g., air temperature, humidity, and rainfall).
- Nonuniform conditions within the solids (e.g., in piles and windrows).

Therefore facilities that use lagoon storage, air drying, and cake storage to produce Class A biosolids must typically monitor viruses and helminth ova, as well as fecal coliform or salmonellae, in each pile or batch of biosolids produced.

From the Class A perspective, a batch is a discrete volume of solids that has undergone the prescribed treatment and has been kept separate from all other solids, including fresh feedstock, during the processing period. Additionally, each increment of solids within the batch has received equal treatment (none have been under-treated). Assuming time is a crucial parameter, the process time clock starts immediately after the batch receives its last charge of fresh feedstock. The process time clock stops when the biosolids are transferred to another (downstream) process, or when the biosolids treatment ends and the final product is sampled to ensure it meets Class A pathogen requirements.

There have been attempts and interest on the part of some operators of these low-technology processes to achieve approval within Part 503 Class A Alternative 6, processes equivalent to a PFRP. The prime sources of this interest are reducing pathogen testing costs and having a defined process that will always produce Class A biosolids assuming certain operating criteria are met. Equivalency reduces the costs and risks inherent in pathogen testing of final products via U.S. EPA Class A Alternative 4 procedures and simultaneously enhances public acceptance of biosolids. Chapter 4.0 describes the process by which equivalency might be obtained.

3.2 Vector Attraction Reduction

Treatment trains operated to produce Class A biosolids (including those that involve lagoon storage, air drying, and cake storage processes) must also satisfy vector attraction reduction (VAR) requirements. Vectors are living organisms capable of transmitting a pathogen from one organism to another either mechanically (by simply transporting the pathogen) or biologically (by playing a specific role in the pathogen's life cycle). Likely vectors for sewage sludge pathogens include insects, rodents, and birds. Aerobic or anaerobic digestion is typically used before lagoon storage, air drying, or cake storage steps to reduce the available food and odor-producing potential that could be attractive to vectors.

The Part 503 regulation requires that most pathogen-reduction processes precede or occur simultaneously with VAR. This requirement came about because Swiss experience showed catastrophic regrowth of salmonellae when pasteurization (a pathogen-reduction process) followed anaerobic digestion (which accomplishes VAR as well as pathogen reduction). The explanation for the regrowth was that pasteurization, in addition to eliminating pathogens, also destroyed nonpathogenic bacteria. These nonpathogenic bacteria compete with pathogens, limiting their growth. The requirement for pathogen reduction before or during VAR ensures the presence of a large number of nonpathogenic organisms during treatment and in the final treated product.

Lagoon storage, air drying, and cake storage processes are typically placed after anaerobic or aerobic digestion. However, VAR begun during digestion continues during these processes and, thus, occurs simultaneously with pathogen reduction. Therefore, the regulation's requirement about the order of process placement is satisfied.

There are 12 options available to satisfy VAR requirements (see Appendix F). Most facilities that anaerobically or aerobically digest their sludge can achieve VAR through Option 1, which requires at least 38% volatile suspended solids (VSS) reduction. VSS destruction during subsequent lagoon storage, air drying, or cake storage processes can be included in Option 1's VSS reduction calculations. VAR requirements can also be met if any one of the following applies:

- Biosolids from the preceding anaerobic digestion system degrade minimally upon further processing (VAR Option 2).
- Biosolids from the preceding aerobic digestion system degrade minimally upon further processing (VAR Option 3).
- Biosolids from the preceding aerobic digestion system display minimal oxygen uptake rates (VAR Option 4).
- The solids content of the product from air drying or cake storage steps exceeds 75% and the product contains no unstabilized solids (VAR Option 7).

Facilities that do not have primary clarifiers (e.g., extended aeration systems) should not count on long aeration times in the wastewater treatment system to produce a product that meets VAR requirements. An additional aerobic digestion step will usually be required unless 75% solids can be obtained in the final product.

CHAPTER 4.0

Process to Further Reduce Pathogens Equivalency

Forty CFR Part 503 Regulations, "Standards for the Use or Disposal of Sewage Sludge," provide six alternatives to meet Class A pathogen requirements. Class A Alternative 5 identifies six specific processes that when operated at or beyond certain conditions produce Class A biosolids. These processes (composting, heat drying, heat treatment, beta-ray irradiation, gamma-ray irradiation, and pasteurization) are designated as PFRPs. If the specified PFRP conditions are satisfied, pathogen-free biosolids will be produced, and the processed biosolids do not have to be analyzed for enteroviruses or viable helminth eggs. Avoiding these analyses is desirable since they are expensive and time consuming. Class A Alternative 6 allows processes to be used to meet Class A requirements that have been determined to be equivalent to PFRPs. If a process has received approval as a PFRP-equivalent process, it too is absolved from the requirement to analyze for enteroviruses or viable helminth eggs. (Some exceptions are noted later in this chapter.)

The regulations identify U.S. EPA regional staff as the authorities who decide whether a process has been demonstrated to be equivalent to a PFRP. U.S. EPA has also established a committee of experts, the Pathogen Equivalency council (PEC), which provides recommendations to the regional regulatory staff.

The PEC has provided guidance on PFRP equivalency. The PEC requests that applicants show the treatment process produces a reduction in enterovirus density of three or more logs (a reduction by a factor of 1000 or more) and a reduction in viable helminth egg density of two or more logs (a reduction by a factor of 100 or more). Methods for measuring densities of these pathogens are provided in the guidance document, "Control of Pathogens and Vector Attraction in Sewage Sludge" (U.S. EPA, 1999). Chapter 11 of the guidance document provides detailed information about the role of the PEC in establishing equivalency. It also provides guidance for organizations that are attempting to demonstrate equivalency for a process. The guidance is generic and has to be adapted to specific situations and processes.

Demonstrating that a process satisfies the aforementioned pathogen-reduction requirements can be easy or difficult. If the process has operating conditions very similar to those of a PFRP named in the regulation or similar to those of an equivalent PFRP, then the applicants may only need to show that their process conditions are at least as severe as those accepted. For a process such as treatment by long-term storage or air-drying, the demonstration is far more complex because these are different processes than existing PFRPs. These processes are also influenced by non-constant, uncontrollable ambient conditions.

Knowledge of factors that affect pathogen reduction is likely to be a requirement for any application. This knowledge helps establish the initial scope of work to demonstrate the process. An equally important requirement is close communication with the PEC throughout the research and demonstration program as well as in the preparation of the formal application for equivalency.

Early in the application process, the proposing organization should provide U.S. EPA (and perhaps the PEC) with an explanation of its process as well as information from the literature on the effect of treatment parameters on process performance. U.S. EPA cannot be expected to be totally familiar with all literature on all potential PFRP-equivalent processes. Early and continuous communication must be maintained with U.S. EPA regarding the experimental program, quality assurance/quality control requirements, and the range of each parameter that should be investigated. The level of confidence in the results that U.S. EPA requires may substantially exceed what the proposing group expects because U.S. EPA would be in an uncomfortable position if it recommended a process as PFRP-equivalent that failed to satisfy pathogen-reduction requirements.

The requirement to work closely with U.S. EPA (and perhaps the PEC) at all stages of the application process can be time-consuming. Communications and applications have to be reviewed by several experts in different locations and a coordinated response must be prepared. This can be a lengthy process. Sometimes extra mandates that result in spending more money and time are required by the PEC. Initial communication with the PEC is important to get an appreciation of the magnitude and standards of the proof that will be required. If costs are more than initially expected, plans for equivalency may have to be revised.

The PEC has provided recommendations on PFRP-equivalency that fall into two categories: national equivalency and site-specific equivalency. A process with national equivalency can be used anywhere in the country. However, qualifications may be included in the approval. For example, for one process that is conducted outdoors, a requirement was attached that the ambient temperature must be above a specified value. The process can be used nationally wherever the temperature requirement is met.

Site-specific recommendations are given when U.S. EPA has reservations that the process may not be applicable to the range of feedstocks or other conditions that could be encountered in the United States. One reason for this limitation might be that the process investigation did not include a broad enough range of variables. Another reason might be that a process may be influenced by ambient temperature and aridity in a way that is not entirely clear. Rather than add a qualification when the effect of such factors is not completely understood, U.S. EPA or the PEC might restrict the approval to the site that is applying for certification. Equivalency has also been provided that is not only site-specific, but requires some monitoring of either viable helminth eggs or enterovirus densities of the final product. This limited approval might be granted when U.S. EPA is not confident that the organism being monitored is consistently destroyed by the process.

For example, the Chicago MWRD applied for equivalency for sludge-treatment trains at its Calumet and Stickney WWTPs. These trains include anaerobic digestion, lagoon storage, and air drying. Virus and helminth densities at these plants are very low, even in the raw (untreated) sludge. The PEC asked the district to conduct large-scale tests to demonstrate 3-log reductions of enterovirus and 2-log reductions of helminth ova. This demonstration would have required spiking the sludge stream with these pathogens. The MWRD felt it was impractical and possibly dangerous to spike viruses and helminth ova into a large-scale test. Instead, it used a statistical approach to estimate helminth and virus inactivation within its sludge-treatment trains in the case of surges in densities of these particular pathogens (Tata et al., 2000a). Eventually the district was awarded site-specific equivalency for its Calumet and Stickney plants. This approval reduced (but did not eliminate) virus and helminth monitoring. Specifically, the MWRD is required to monitor once per month for enteric virus and helminth ova for the first year after receiving certification and once every other month after that, as long as it operates its sludge treatment processes according to the process and operating protocols specified in its application (Traub, 2002). This is the only known U.S. EPA approval for PFRP equivalency for sludge-treatment trains involving these low-technology processes.

Probably the biggest single problem in satisfying U.S. EPA requirements has been the difficulty in meeting the PEC's desire to confirm the demonstration of desired pathogen reductions on a large scale. The PEC is aware that a full-scale plant is not necessarily a duplicate of pilot-scale demonstration plant, so it is wary of making equivalency decisions based on pilot-scale demonstrations. This conflicts sharply with municipalities' reluctance to introduce increased quantities of pathogenic microorganisms into their wastewater solids systems. Demonstrating the desired log reductions based on the numbers of pathogens ordinarily found in wastewater solids streams can be an extremely difficult task, and probably impossible without enormous time and money expenditures. Satisfactory compromises have not been reached between the PEC and municipalities on this sensitive subject. Groups that want to demonstrate a proposed PFRP-equivalent process should address this issue at the outset. Creative thinking may find solutions.

CHAPTER 5.0

Class A Processing Conditions for Lagoon Storage, Air Drying, and Cake Storage Systems

Three factors affect the ability of low-technology sludge treatment trains to produce Class A biosolids:

- *Pathogen densities in the raw (untreated) sludge.* Densities can differ greatly by location. For example, Reimers et al. (2001) reported measurements of viable helminth ova in raw sludge from 14 large cities. Mean values ranged from 0.8 to 136 eggs/4 g dry solids.
- *Performance of sludge stabilization units in the process train other than the low-technology processes* (e.g., aerobic or anaerobic digestion)
- *Performance of the low-technology process(es)* (lagoon storage, air drying, and/or cake storage)

This chapter addresses the third factor, performance of low-technology sludge treatment processes. However, the other two factors must be considered when developing the total treatment train.

5.1 The Ideal Low-technology Treatment Process

The product from an *ideal* low-technology sludge treatment process, or combination of processes, can satisfy pathogen standards for Class A biosolids and meet the criteria for a process equivalent to a PFRP. Fulfillment of the following five requirements can help achieve process objectives:

1. Processes that precede the low-technology process(es) must significantly reduce pathogen densities and the bacterial food supply. They must also develop and sustain a healthy population of non-pathogenic bacteria. These actions will help ensure high pathogen kill (Requirement 2), satisfy VAR requirements (Requirement 3) and help prevent bacterial regrowth (Requirement 4).
2. The treatment process(es) must be of sufficient intensity and duration to ensure requisite pathogen kill. Helminth ova are typically more resistant to treatment than other regulated organisms (fecal coliform, salmonellae, and viruses). Providing the conditions needed to inactivate helminth ova should ensure satisfaction of Class A criteria for these other organisms.
3. The overall treatment train must satisfy VAR needs.
4. Bacterial regrowth must be prevented.
5. Treatment must be uniform to ensure that all biosolids have been properly treated.

The next section provides guidelines on how to fulfill these requirements.

5.2 Process Guidelines

The guidelines presented here were developed and detailed in Appendix G, and should be used with discretion and caution. They are empirical, evolving from observations made under a variety of conditions including laboratory-, pilot-, and full-scale operations. Because the relative importance of mechanisms responsible for pathogen kill in lagoon storage, air drying, and cake storage processes are not completely understood and are therefore difficult to quantify, because feedstock quality varies by location, and because plant operators may not use identical operating procedures, results obtained at one location may not transfer well to another. Consequently, the reader must never assume that using these guidelines ensures production of a Class A product. Pathogen monitoring is required to gauge process performance. Nevertheless, the guidelines should help those considering the use of these processes to make reasonable design and operating decisions.

5.2.1 Lagoon Storage

Few plants produce Class A biosolids with lagoon storage as the only low-technology sludge treatment process. For Class A applications, lagoons are often used in combination with air drying. This combination provides more time and different mechanisms for pathogen destruction, and allows the processed material to be distributed as a dewatered or dry product. However, application of liquid biosolids on farm and forest lands is practiced. In some instances, application of slurry-type Class A biosolids (as opposed to slurry-type Class B products) on land may have certain advantages. For example, setbacks from the property line are required when Class B biosolids are land applied. If Class A biosolids are land-applied, no setbacks are needed. Therefore, there may be a niche for distribution of Class A biosolids from lagoon storage.

An analysis of the information from this project's literature review suggests that Class A biosolids may be produced by lagoon storage under the following conditions:

- Requirement 1 (providing adequate sludge pretreatment) is satisfied by preceding the process with aerobic or anaerobic digestion. Pathogen reduction proceeds more rapidly in downstream processes if anaerobic digestion pretreatment is used (Kaneshiro and Stern, 1985; Reimers et al., 2001). When mesophilic anaerobic digestion is used, solids retention times (SRTs) of at least 15 days are preferred.
- Requirement 2 (providing requisite helminth kill) is satisfied by providing enough time to reduce the density of viable helminth ova to the Class A limit (1 viable ovum/4 g dry solids). Biosolids should be held in the storage lagoons for a certain amount of time (called "hold time" or "unfed storage time") following the last biosolids addition.
- Data presented in Appendix G show that laboratory and pilot-scale storage systems operating at 25°C (77°F) reduced the density of viable *Ascaris* ova (the most-resistant pathogen) in spiked, anaerobically digested biosolids to below the Class A criterion within five to 16 months. Viable *Ascaris* ova were reduced significantly (3.7 to 4.4 log reduction) during this process. Therefore lagoons operating at ≥25°C (77°F) appear capable of producing Class A biosolids within 16 months under most conditions, and probably well under 16 months with typical conditions.

Appendix G data also show that storage of spiked, digested biosolids at 4°C (39°F) for 25 months reduced the density of viable *Ascaris* ova to a much lesser extent than at 25°C (77°F) (0.8 to 1.6 log reduction), insufficient to achieve the U.S. EPA Class A criterion. However, if the log reduction performance found in these laboratory-scale experiments can be maintained with the much lower *Ascaris* densities that are typically found in municipal sludge (and there is no certainty that it can be), then lagoons operating at 4°C (39°F) might produce Class A biosolids within 25 months, provided the densities of viable helminth ova in lagoon feed are relatively low (≤10 ova/4g dry solids).

There is less information about the rate of inactivation of the *Ascaris* ova at temperatures between 4°C and 25°C (39°F and 77°F). Therefore, definitive "hold time" guidance cannot be given for this temperature range. Such information is a future research need, since most storage lagoons operate in this range.

- Requirement 3 (meeting VAR objectives) is typically fulfilled within time frames needed to satisfy *Ascaris* kill through VAR Options 1, 2, 3, or 4.
- Requirement 4 (preventing bacterial regrowth) can be satisfied by providing unfed storage time long enough to ensure requisite helminth kill. During this time the bacterial food supply will be reduced to levels insufficient to sustain bacterial regrowth.
- Requirement 5 (ensuring uniform treatment) is accomplished well in long-term lagoon storage systems.

5.2.2 Air Drying or Cake Storage

This project's literature review suggests that Class A biosolids might be produced by air drying or cake storage when the following requirements are satisfied:

- Requirement 1 (providing adequate sludge pretreatment) is satisfied by preceding either process with aerobic or anaerobic digestion. When mesophilic anaerobic digestion is used, SRTs of at least 15 days are preferred.
- Requirement 2 (providing requisite helminth kill) may be satisfied by desiccation, time, and temperature.
  - *Desiccation-oriented air drying systems.* It is necessary to reduce biosolids moisture contents to below 5% to achieve complete inactivation of helminth ova by desiccation. Operators of most U.S. air drying plants will be unable to dry their biosolids to moisture levels of ≤5% because of rainfall or humidity constraints. Operators in hot, arid regions (e.g., the southwestern U.S.) may be the exception. Even those who can achieve such low moisture levels may be unwilling to do so because of dusting problems. Drying time and drying temperature still play roles in desiccation-oriented drying systems. However, time and temperature guidance cannot be given for these systems at this time.
  - *Time- and temperature-dependent systems.* Air drying systems that do not rely on desiccation (the great majority) and cake storage systems must provide processing times long enough and operating temperatures high enough to ensure the requisite destruction of helminth ova. These air drying or cake storage systems should be operated for at least 250 days in locations with warm summers and mild winters and at least 350 days in locations with warm summers and cold winters (See Appendix G for the basis for these drying/storage times). A summer should be included in the

operating period. Pile or windrow moisture content should remain above approximately 40% for most of the storage or drying periods to promote continued biological activity. Windrows should be turned to control the windrow moisture content and temperature.

- Requirement 3 (meeting VAR objectives) can be satisfied through VAR Option 7 for desiccation-based air drying systems (stabilized biosolids with moisture content <25% satisfy Option 7 as long as they are not rewetted). For systems that do not dry to low moisture contents, VAR objectives could be satisfied within the time frames necessary to satisfy helminth kill requirements (250 to 350 days) through compliance with Options 1, 2, 3, or 4 of the VAR regulations. The pile or windrow should remain moist enough (>40% moisture) for most of this period to promote continued biological activity.
- Requirement 4 (preventing bacterial regrowth) may be problematic for desiccation-based air drying systems. If drying has been rapid and insufficient organic matter has been destroyed, regrowth may occur if the biosolids are rewetted. Systems that rely primarily on time and temperature to achieve the requisite pathogen kill may experience fewer regrowth problems. They will be operated long enough (250 to 350 days) that residual organic matter should be insufficient to support bacterial regrowth.
- Requirement 5 (ensuring uniform treatment) can be achieved in windrow systems by periodically turning the sludge. Turning also promotes autoheating, which encourages drying.

5.2.3 Coupled Lagoon Storage/Air Drying Systems

As mentioned earlier in this chapter, lagoon storage is often used in combination with air drying for Class A applications. Because organisms exit the lagoons in a debilitated state, they are vulnerable to the altogether different environment they encounter in the air drying process. This sequential combination is probably more powerful than lagoon storage or air drying alone, even when temperature conditions are the same for all. Therefore, guidelines for sequential lagoon storage and air drying systems are different from guidelines presented previously, which were for individual systems.

- Requirement 1 (providing adequate sludge pretreatment) is satisfied by preceding either process with aerobic or anaerobic digestion, with the latter preferred. SRTs for mesophilic anaerobic digesters should be at least 15 days.
- Requirement 2 (providing requisite helminth kill) is satisfied by providing processing times long enough and operating temperatures high enough to ensure requisite destruction of helminth ova. The Chicago MWRD treatment train serves as a model for combined lagoon storage and air drying systems. The district's storage lagoons operate with a minimum of 18 months unfed storage time. The downstream air drying systems are operated to produce a 60% solids product within 9 to 10 weeks during warm weather. Air drying windrows are turned three times a week. Treatment times for other coupled systems may be less in warmer climates, and more in colder climates.
- Requirement 3 (meeting VAR objectives) is typically fulfilled within time frames necessary to satisfy helminth kill through compliance with Options 1, 2, 3, or 4 of the VAR regulations.

- Requirement 4 (preventing bacterial regrowth) can be achieved by employing adequate sludge pre-treatment (digestion) and by operating the combined systems long enough to ensure helminth kill. These operating conditions will reduce the bacterial food supply to such low levels that bacterial regrowth cannot be supported.
- Requirement 5 (ensuring uniform treatment) is achieved well within long-term lagoon storage systems coupled with air drying, especially when air drying includes periodic turning of the biosolids.

CHAPTER 6.0

RESEARCH NEEDS

The following is an outline of recommended research tasks:

- *Develop practical methods for assessing helminth kills in full-scale lagoon storage, air drying, and cake storage systems and the time/temperature conditions necessary to achieve these kills.* While there is a fair amount of information about the densities of helminth ova in the products from such systems, there are relatively few data about the densities of helminth ova entering them, and thus relatively little information demonstrating the treatment effect. Information about helminth kills in lagoons is particularly lacking in the temperature range of 10°C to 20°C (50°F to 68°F). This task is especially difficult due to the low densities of helminth ova typically in the feed streams of municipal sludge treatment systems. One alternative may be to use small chambers containing biosolids seeded with known densities of viable helminth ova. These chambers are suspended within the biomass of the full-scale sludge treatment systems. The chambers are configured so the environment within the chambers is the same as within the full-scale process. The decline of viable ova is then tracked by analyzing ova densities within the chambers over time. This approach could improve experimental accuracy while simultaneously reducing experimental complexity and costs. Experimental protocols must be developed so that organisms within the chambers are truly subjected to the conditions they would experience in full-scale treatment systems.
- *Determine temperature/moisture profiles in biosolids windrows, drying beds, and cake piles.* This would lead to greater understanding of spatial variability in treatment results.
- *Develop conclusive data that viruses in municipal sludge die off faster than viable helminth ova.* Poliovirus and bacteriophage f-2 (an indicator organism for enteroviruses) die off much more rapidly than viable helminth ova in cake storage systems that generate finished product with moderate moisture levels (25% to 50% moisture) (Ahmed and Sorensen, 1995; Ahmed and Sorensen, 1997). Reimers et al., 2001 indicate that poliovirus die off more quickly than helminth ova in lagoon systems. If data could be developed which showed that all or the large majority of viruses found in municipal sludge died off faster than helminth ova, regulators might be convinced to eliminate the virus monitoring requirement for Alternative 4 systems. The absence of helminth ova would then be sufficient evidence that virus are absent.
- *Develop a consensus for an improved and more-accepted assay for quantifying viable and nonviable helminth ova in biosolids.* Since survival of helminth ova controls treatment time for low-technology processes, there is a great need for standardization of helminth ova analyses. Bowman, et al. 2003 cited accuracy and precision data from studies conducted using the Tulane method for measuring the concentrations of viable ova of *Ascaris suum.* They were able to recover 70% to 80% of spiked viable ova from lagoon-stored biosolids with a precision (differences between duplicate samples) of 10% to 15%. The recovery rates were considerably higher than recovery rates reported elsewhere. The authors suggested that U.S. EPA's analytical method (1999) has several shortcomings, including too high a ratio of solids to dilution water in the initial sample dilution and an incomplete and ambiguous procedure for examination of the cultured sediment with a Sedgwick-Rafter cell. They noted

that recovery and precision varied with the type of biosolids being examined and that the Tulane procedure, which works well for *Ascaris* ova, may need to be modified to achieve equivalent results with other types of helminth ova. They noted the need for preservation and storage methods to quench the inactivation process and reduce natural metabolic inactivation during the interval between sample collection and the actual analytical measurement.

- *Evaluate different approaches to increase the concentration of un-ionized ammonia in systems designed to produce Class A biosolids, including adjusting pH and possibly adding urea or other compounds.* Un-ionized ammonia (NH_3) is a virucide and an ovacide of some potency, but is present in relatively low concentrations at the pH of lagoon storage, air drying, and cake storage operations (pH 6.0 to 7.0). Raising biosolids pH moderately (one to two units) increases the biosolids NH_3 concentration dramatically. However, the effects of moderate pH increases are not fully understood, and it is uncertain if this approach would be cost effective.
- *Develop a practical method for determining the potential for regrowth of fecal coliform or salmonellae in stabilized biosolids product.* This research topic is not limited to low-technology Class A processes. It is also important for other conventional and new biosolids treatment processes. It is of great importance for Class A processes due to concerns for the public to accept use of the material.

APPENDIX A

REVIEW OF THE LAGOON STORAGE LITERATURE

The literature the project team reviewed concerning pathogen die-off in lagoon storage of municipal biosolids is summarized below.

A.1 Pedersen (1981)

Pedersen (1981) conducted a comprehensive review of literature published between 1940 and 1980 from laboratory and full-scale studies regarding density levels of indicator and pathogenic organisms in municipal wastewater sludge and septage, and the effects of all types of processing on these levels. Effects of drying beds and lagoons on density levels in biosolids are included. The author reviewed the relatively small database on the effect of lagooning on microorganisms in lagooned biosolids. She concluded that *Ascaris* ova is the critical organism in lagoon storage, requiring at least six months storage if temperatures are above 20°C (68°F) and over three years storage if temperatures are below 20°C (68°F).

A.2 Bitton et al. (1980)

Data by Bitton et al. (1980) are cited by Pedersen as showing the decline of enteric viruses in a lagoon in Florida that received no new biosolids for six months. Densities were reduced by about 3 logs in two months (April and May). Farrah et al. (1981) noted that solids in a north Florida lagoon no longer contained detectable viral infectivity 1.5 months after the addition of fresh biosolids was halted.

A.3 Sattar and Westwood (1979)

Sattar and Westwood (1979) conducted a 15-month field study on anaerobically digested biosolids in a drying lagoon in Ottawa, Canada. Their studies showed complete reduction of viruses within eight months of retention time after biosolids feed was terminated. Samples collected up to 13 months after ceasing biosolids feed were negative for viruses. However, the authors stated that these results should not be taken to indicate the disappearance of viruses from the biosolids because of the limited number of samples taken during months nine though 14 (the lagoon surface was frozen and covered with snow for much of that time). Such conditions are more conducive to preserving infectivity of viruses than to their inactivation.

A.4 Veerannan (1977)

In a laboratory study with seeded digested biosolids, Veerannan (1977) found that 50% of seeded *Ascaris lumbricoides* were still viable after one year, 25% after 1.5 years, 12% after two years, and that practically none survived after three years of anaerobic storage. The storage temperature was not specified.

A.5 Stern and Farrell (1977)

Stern and Farrell (1977) at U.S. EPA's laboratory in Cincinnati reported results of a limited investigation of the effects of storage on microorganism kill under controlled conditions. They stored digested primary sludge at two temperatures (4°C and 20°C [39°F and 68°F]) for 24 weeks, and observed the decline in fecal coliform, salmonellae, and enteroviruses. At 4°C (39°F) in six months fecal coliform densities dropped almost 5 logs, salmonellae dropped 2 logs, and enteroviruses fell about 0.5 log. At 20°C (68°F), densities of all of these microorganisms fell much more rapidly. Fecal coliform densities dropped 3 logs in eight weeks, and salmonellae and enteroviruses fell more than 2 logs. A qualitative examination showed that helminth ova had survived for 24 weeks even at 20°C (68°F).

A.6 Kaneshiro and Stern (1985)

The results of the above bench-scale experiments by Stern and Farrell (1977) were sufficiently encouraging to justify a much more extensive investigation of survival of microorganisms in long-term lagoon storage. Kaneshiro and Stern (1985), at the University of Cincinnati, instituted a comprehensive laboratory-scale investigation of the survival of ova of several types of helminth. Viable helminth ova were added to aerobically and anaerobically digested biosolids (2.0% to 3.5% solids). A spike of the several ova types was added before the solids were digested as well as after digestion. All ova examined were obtained from fecal material. Two temperatures were chosen, 4°C and 25° C (39°F and 77°F). The hardiest helminth ova were *Ascaris suum* (the roundworm of pigs, whose ova are indistinguishable from the ova of *A. lumbricoides,* the human roundworm, and are considered to have the same resistance to environmental exposure). They survived in substantial densities at 4°C (39°F), even after 25 months. At 25°C (77°F), the density of viable ova were reduced to negligible values after 16 months (5-log reduction), and the ova were non-infective after that time. There was little difference in the long-term performance regardless of how biosolids had been digested.

The authors also stored the various biosolids underground to simulate the performance of a lagoon in Cincinnati's climate. Here, the pretreatment method made a difference. After 16 months of storage, viable *Ascaris* ova were virtually eliminated (5-log density reduction) if the biosolids had been anaerobically digested. If they had been aerobically digested there was substantial survival. However, the ova in the anaerobically digested biosolids were still infective after up to 22 months of storage. Average temperature for the 16-month period was 13.8°C (57°F), with a range from -5°C to 28°C (23-82°F).

The disappearance of the ova is the result of two effects: the reduction in the total number of ova and the reduction in viability. Various researchers have theorized that much of the reduction in density is the result of the action of bacteria or fungi in which enzymes released cause disintegration of the ova. The reduction in viability could be caused by the same mechanism or by penetration of the ova's coat by chemicals toxic to ova, such as ammonia.

A.7 OíDonnell et al. (1984)

O'Donnell et al. (1984) confirmed Kaneshiro and Stern's results. They reported that *Ascaris* and *Toxocara* eggs seeded into biosolids before and after anaerobic and aerobic digestion retained viability and infectivity for as long as two years when stored at 4°C (39°F), but were destroyed in 10-16 months of storage at a temperature of 25°C (77°F).

A.8 Reimers et al. (1989, 2001)

Tulane University conducted a pilot-scale demonstration of lagoon stabilization (Reimers et al.1989 and 2001). The lagoons, 1.8 m x 1.8 m x 3.0 m deep, (5.9 ft x 5.9 ft x 9.8 ft deep), contained about 7.5 m^3 (2000 gallons) of anaerobically digested biosolids and were constructed in New Orleans, Louisiana, and Port Aransas, Texas. The entire mass of biosolids in the lagoons was spiked with salmonellae, poliovirus, and *A. suum* collected from pig feces.

Water temperatures varied from 20-29°C (68-84°F) from April to October and 12-24°C (54-75°F) from November to March. Reduction of pathogens and indicator organisms are shown in Table A-1.

The results for *Ascaris* were confirmed by repeats of the experiment with different batches of the biosolids collected from the same treatment plants but at a later time, using tubes instead of pilot lagoons. The tubes were about 2.4 m (8 ft) in height and 0.10 m (4") in diameter. They were submerged in the lagoons so that they obtained the same temperatures as the lagoons. They produced the same reductions in *Ascaris* densities as the in-ground pilot-scale lagoons.

The authors also conducted tests with the same biosolids stored in two large 3.66 m (12 ft) long by 0.3 m (1 ft) diameter columns in New Orleans. The column tests were conducted because the authors feared that previous tests had been made with ova of low viability. The *Ascaris* ova spiked into the larger columns had a higher percentage of viability than those used in the previous tests. Complete ova inactivation in the column with New Orleans biosolids took three months longer than ova inactivation in the New Orleans lagoons and tubes. Complete ova inactivation in the column with Port Aransas biosolids took six months longer than ova inactivation in the Port Aransas lagoons and tubes.

Table A-1. Pathogen Inactivation with Time for Lagoon-stored Anaerobically Stabilized Biosolids.

Organism	Time for log reduction (months)	Log reduction
Total coliforms	9	2
Fecal coliforms	6	>4
Fecal streptococcal organisms	3	1 to 3
Salmonellae	2	>7
Poliovirus	6	>5
Ascaris eggs		
New Orleans	12	2.0
Port Aransas	6	2.5

Source: Reimers et al. (2001)

Reimers et al. (2001) noted that pathogens tend to die off faster in a biosolids matrix than in soils, speculating that this result might be due to the higher content of organic acids, aldehydes, alcohols, and ammonia in the biosolids matrix as well as higher densities of enzymes and biological activity of all types. They also noted that die-off rates were greater in the Port Aransas study, attributing that result to petroleum contamination. They also cited studies at Tulane University (Mbela, 1989) that showed that within the 25-45°C (77-113°F) temperature range, anaerobic digestion inactivated *Ascaris* eggs 40% faster than aerobic digestion. Again, they cited the higher content of organic acids, aldehydes, alcohols, and ammonia in the anaerobic system as the reason for its greater effectiveness. This result suggests that coupled anaerobic digestion/lagoon storage systems might be more effective than coupled aerobic digestion/lagoon systems in inactivating *Ascaris* ova.

Reimers et al. (1985) studied inactivation of the eggs of the parasites *Ascaris, Toxacara, T. trichuria,* and *T. vulpis* in two biosolids storage lagoons in Minnesota. The lagoons received anaerobically digested biosolids. The lagoons had minimal treatment effect in the winter. In the summer and fall, the lagoons appeared to inactivate the eggs of *Ascaris, T. trichuria,* and *T. vulpis.*

A.9 Ponugoti et al. (1997)

Ponugoti et al. (1997) investigated the fates of fecal coliforms, fecal streptococci, and *Salmonella sp.* in the sludge-processing systems of 12 WWTPs in central and eastern Nebraska. One plant stored aerobically digested biosolids in a lagoon for approximately six months. Storage reduced fecal coliform densities from 3.8 million to 0.2 million MPN/g total solids; fecal streptococci densities from 2.23 million to 90,000 MPN/g total solids; and *Salmonella sp.* densities from 2340 to 256 MPN/4 g total solids. The storage process failed to satisfy Class A criteria for fecal coliforms and *Salmonella sp.* The authors provided no temperature data. It is also not clear if the lagoon was fed fresh solids during the storage period.

A.10 McHaney et al. (1994)

McHaney et al. (1994) reported on results from the solids stabilization system of the San Jose/Santa Clara, California, Water Pollution Control Plant. The stabilization system combines anaerobic digestion, lagoon storage, and air drying processes. This appendix reports on the lagoon storage process. Appendix B will describe air drying operations and Appendix D their combined lagoon storage/air drying process.

Raw sludge and thickened waste sludge from the suspended growth secondary and nitrification units are anaerobically digested at 35°C (95°F) for up to 40 days. The digested biosolids (1.5% solids) are pumped into about 30 10-ft deep, 20-million gallon storage lagoons. Lagoon supernatant is returned to the plant headworks. Lagoon water temperatures vary from 15°C to 18°C (59°F to 64°F). An analysis from March 1994 showed lagooning reduced coliforms greatly (from 3.9×10^6 to 129 MPN/g dry solids) and salmonellae negligibly (from 167 to 159 MPN/4 g dry solids). The lagoon product satisfied the Class A criterion for fecal coliform but not the Class A criterion for salmonellae. At the time of the analysis, San Jose did not operate the lagoons with storage times without feed. Therefore, pathogen reduction was not as good as in later years when a two-year unfed storage time was instituted.

A.11 Tata et al. (2000); Lue-Hing et al. (1999)

Tata et al. (2000) reported on the results of three-year, full-scale studies at Chicago MWRD's Calumet and Stickney wastewater reclamation plants. Each plant uses two similar but distinct solids processing trains (SPTs). Anaerobic digestion (35°C [95°F], 20 days SRT) is the first step in the SPT. Anaerobically digested biosolids are further processed by either of two methods:

- *High-solids SPT (HSSPT).* The anaerobically digested biosolids are dewatered to a 25% solids cake with centrifuges. The cake is then transported by rail to lagoons where it is stored and aged to further inactivate pathogens for at least 18 months (18 months minimum unfed storage time in lagoons). After storage and stabilization in the lagoons, the cake is removed from the lagoons and air dried to approximately 60% solids on paved beds.
- *Low-solids SPT (LSSPT).* The anaerobically digested biosolids are transferred to lagoons which are filled in cycles over one to three years. Batches are added layer-by-layer until the biosolids fill the lagoon, at which point the biosolids have concentrated to 15% solids. After storage and stabilization in the lagoons (at least 18 months of unfed storage time), the biosolids are removed from the lagoons and air dried to approximately 60% solids on paved beds.

Previously, the MWRD had operated the SPTs to produce a 60% solids air-dried product for beneficial use. Pathogen reduction had not been the primary goal. In the study reported, the district optimized and codified SPT protocols and imposed additional controls to ensure the modified SPTs would meet Class A criteria.

This appendix discusses the district's lagoon operations. The district's air drying operations are discussed in Appendix B and overall system operations in Appendix D.

Minimum solids holding times for HSSPT and LSSPT lagoons were set at 1.5 years after last placement of biosolids. Actual holding times were longer. The average holding times for Calumet and Stickney LSSPT lagoons were 2.5 and 2.0 years, respectively. The average detention times for Calumet and Stickney HSSPT lagoons were 1.6 years. Measured sludge layer temperatures are shown in Table A-2.

Table A-2. Sludge Layer Temperatures (°C) in the Calumet and Stickney Lagoons.

	Calumet		Stickney	
Item	**LSSPT**	**HSSPT**	**LSSPT**	**HSSPT**
Low temperature	12	7	6	6
High temperature	24	21	21	20
Mean temperature	15	13	13	14

Source: Lue-Hing et al. (1998)

Digester feed, digester draw, lagoon draw, and air-dried product were analyzed for fecal coliforms, salmonellae, viruses, and viable helminth ova. Table A-3 presents results for reduction of these organisms in the lagoons.

Digester draw and lagoon draw virus and helminth ova densities were very low and within Class A requirements. Therefore it was difficult to directly demonstrate that lagoon storage produced high log reductions of these organisms.

Parallel to full-scale operations, Chicago conducted studies in storage basins of at least 153 m^3 (200 yd^3) volume at the Stickney and Calumet plants (Lue-Hing et al., 1998). At each plant, HSSPT pilot basins were filled with centrifuge cake, and LSSPT pilot basins were filled with digester draw. The pilot basins were completely filled within 72 hours and no biosolids were added to them for the duration of the tests. The pilot basins were sampled 24 hours after filling and then at 1, 2, 4, 6, 15, and 18 months. The purpose of the pilot studies was to determine how long after filling it would take the density of each pathogen to reach Class A limits.

The HSSPT pilot basins more effectively inactivated fecal coliforms than the LSSPT basins. Fecal coliforms were reduced to the Class A limit in about two months in Stickney and Calumet HSSPT and LSSPT pilot basins. There was evidence of some regrowth in the Calumet LSSPT basin. Salmonellae were reduced to the Class A limit in about four months in all systems except the Calumet LSSPT pilot basin, where salmonellae densities remained poised at or below the limit for the whole test. Virus and helminth ova densities were low initially and subsequent samples were at or below the Class A limit for both species.

Table A-3. Reduction of Pathogens and Pathogen Indicators in Calumet and Stickney Lagoons.

System	Organism	Pathogen Density		
		Digester Draw	Lagoon Draw (codified)[a]	Log_{10} Reduction
Calumet HSSTP	Fecal coliforms, MPN/g DS[b,c]	6.78 x 10^4	105	2.81
	Salmonella sp, MPN/4 g DS[d]	74.6	0.828	1.95
	Virus, PFU/4 g DS[d]	0.428	0.00	--
	Viable helminth ova, no./4 g DS[d]	0.286	7.44 x 10^{-3}	1.58
Calumet LSSTP	Fecal coliforms, MPN/g DS	6.78 x 10^4	68.5	2.99
	Salmonella sp, MPN/4 g DS	74.6	2.9	1.41
	Virus, PFU/4 g DS	0.428	0.00	--
	Viable helminth ova, no./4 g DS	0.286	0.0625	0.66
Stickney HSSPT	Fecal coliforms, MPN/g DS	2.02 x 10^5	188	3.03
	Salmonella sp, MPN/4 g DS	25.7	0.276	1.96
	Virus, PFU/4 g DS	0.121	0.00	--
	Viable helminth ova, no./4 g DS	0.379	0.784	(0.31)
Stickney LSSPT	Fecal coliforms, MPN/g DS	2.02 x 10^5	151	3.13
	Salmonella sp, MPN/4 g DS	25.7	0.533	1.68
	Virus, PFU/4 g DS	0.121	0.00	--
	Viable helminth ova, no./4 g DS	0.379	0.0394	0.98

[a] Codified means operating at conditions as stringent or more stringent than specified by the operating protocols.
[b] Geometric mean
[c] DS = dry solids
[d] Arithmetic mean
Source: Tata et al. (2000)

Reimers et al. (2001) analyzed results of Chicago's cake storage operations and concluded that the HSSPT storage systems reduced the densities of pathogens and indicator organisms more efficiently than the LSSPT storage systems.

Bergman and Rohloff (1996) reported costs for Chicago's sludge-treatment operations in the years 1992 through 1995. LSSPT lagoon dewatering and outloading costs averaged \$11.08 and \$14.09 per ton dry solids (\$12.18 and \$15.50 per tonne dry solids), respectively. HSSPT lagoon conditioning and outloading costs averaged \$0.44 and \$10.00 per ton dry solids (\$0.48 and \$11.00 per tonne dry solids), respectively. Centrifuging costs were not included in the HSSPT cost calculations.

APPENDIX B

REVIEW OF THE AIR DRYING LITERATURE

The literature the project team reviewed concerning air drying of municipal biosolids is summarized below.

B.1 Pedersen (1981)

Pedersen (1981) reviewed the results of several studies on the survival of helminth ova in air-dried biosolids. She concluded that moisture contents <5% appear to be lethal to *Ascaris* ova, citing Cram (1943; see Section B.5) among others. The data also show regrowth potential for bacteria in drying beds (Baxter et al. 1983; see Section B.6), and a rapid increase in the rate of destruction of enteric viruses when the solids content of drying biosolids exceeds 60% solids (Ward and Ashley, 1977; see Section B.4). The bacteriophage f-2, considered to be hardier than the enteroviruses to environmental stress (Baxter et al. 1983), was shown to be more susceptible to air drying than fecal coliform.

B.2 Reimers et al. (1986)

Reimers et al. (1986) quantified densities of helminth ova in the biosolids from a number of plants in several northern states (Washington, New York, Ohio, and Minnesota). Densities were lower than in biosolids from southern states (Louisiana, Alabama, Texas, Mississippi, and Florida). Densities from the plants examined in Washington were higher than in plants from Ohio, New York, and Minnesota. The authors conclude that densities of viable helminth ova in biosolids that had been applied to drying beds decreased with decreasing moisture of the dried product. They attempted to correlate results with age on the bed, temperature, type of digestion, climate, and moisture content but could not establish any statistically significant relationships.

B.3 Amin (1988)

Amin (1988) reported on the density of pathogens present in dried raw sludge on the middle-eastern island of Bahrain. Raw primary sludge and secondary sludge from the Tubli Water Pollution Control Center are thickened then spread on uncovered drying beds. Bahrain's high humidity does not permit extensive drying. Dried sludge may contain 30-40% moisture. Ambient temperatures range from 14°-40°C (57-104°F).

Poliomyelitis type 2 and Coxsackie type B-5 were not detected after one year's storage. Total coliforms were measured at 5400 per g, while *E. coli* ranged from 15 per 50 g to 3500 per g. *Salmonella* sp. and *Shigella* sp. were detected, but not quantified. Fecal streptococci ranged from 9 per g to 1750 per g. The mean density of helminth ova were 21 per 10 g and 1 per 10 g for *Ascaris lumbricoides* and *Hymenolepis nana*, respectively. The mean density of larvae and adults of the helminth species *Strongyloides sterocoralis* was 3 per 10 grams. The author concluded that the dried sludge posed a potential health hazard.

B.4 Ward and Ashley (1977)

Ward and Ashley (1977) conducted experiments to demonstrate the effect of drying on poliovirus inactivation. They showed that inactivation of poliovirus accelerated rapidly as the sludge moisture content dropped into the range of 35% to 17%. To eliminate the inactivation effects of temperature, they accomplished drying via evaporation at a constant temperature of 21°C (70°F). To eliminate the virucidal effect of ammonia, drying was conducted in ammonia-poor raw sludge samples instead of ammonia-rich digested sludge samples. The loss of infectivity during drying was deemed due to irreversible inactivation of poliovirus because virus particles were found to have released their RNA molecules, which were extensively degraded. Inactivation appeared to be mainly through the action of evaporation, rather than the effect of other factors (e.g., increasing surface to volume ratio of the sludge sample or increasing concentration of virucidal components).

The authors also found that evaporation inactivated coxsackievirus B1 and reovirus type 3 when the moisture content decreased to around 7% and 6% to 27%, respectively. Coxsackievirus B1 is known to be resistant to high pH and reovirus type 3 is resistant to high temperature and ammonia. Therefore, drying was evaluated as a potentially effective means of inactivating these viruses. (The authors suggest that most enteroviruses besides reoviruses are sensitive to ammonia.)

B.5 Cram (1943)

Cram (1943) found *Ascaris* eggs in anaerobically digested biosolids survived long drying periods. They were viable for as long as 118 days of indoor (greenhouse) drying, in which climate was controlled, and 170 days of outdoor drying, in which climate was not controlled but air movement and sunlight were present. The eggs were resistant to moisture loss. Viable eggs were found in cakes with moisture contents <10%. One lot of biosolids cake contained viable eggs when its moisture content was 5.8% after 81 days of greenhouse drying with summer temperatures, which frequently reached 46°C (115°F) in the greenhouse. However, only disintegrated *Ascaris* eggs were found in cakes with 3% and 4% moisture. Hookworm eggs developed and active larvae hatched, continuing development up to the sheathed infective stage for periods of up to 62 days and until the moisture content dropped to as low as 9.9%.

B.6 Baxter et al. (1983)

The results of Baxter et al. (1983) show excellent reduction of all microorganisms of concern in Denver's climate. They found that fecal coliforms were reduced to low densities and salmonellae were eliminated in 15 weeks or more of air drying of biosolids in earthen basins. Moisture content was reduced to <5% in all cases, but in one case there was a regain in moisture when the drying period extended into the winter. In one out of four cases there was evidence of very limited regrowth of both types of bacteria after about 20 weeks that persisted for about three weeks. Regrowth can occur when there is sufficient food available for the microorganisms (organic matter has been insufficiently destroyed) and the biosolids are rewetted. Reduction in *Ascaris* densities could not be followed because initial densities in the biosolids were too low. Viral reductions were measured by use of the bacteriophage f-2 which is hardier than enteroviruses under environmental exposure. In all cases, f-2 densities showed greater log reductions than fecal coliform for a given drying exposure.

In one experiment, a portion of the biosolids was spiked with *Ascaris* ova and their decline measured. Densities of viable ova declined steadily as the biosolids dried. After about 20 weeks, moisture content was below 7.7%; total numbers of identifiable ova were <10% of the original density and only 4% were viable. Viability was zero when moisture content fell below 5%. The temperature of the drying biosolids was not recorded. Since this experiment ended in late August, temperature of the drying biosolids could have been relatively high. Consequently, it is not certain whether the destruction of *Ascaris* ova was due to elevated temperature or desiccation or both.

B.7 Wilson and Co. (1990)

An investigation conducted by Wilson and Co. (1990) at Roswell, New Mexico, sheds light on some of the factors that affect bacterial reduction in air drying of biosolids. In their study, drying rates of anaerobically digested biosolids on conventional sand drying beds were compared with drying rates on paved beds. The biosolids on sand beds were applied to a depth of about 9 in. and allowed to dry. On the paved beds, the biosolids were applied to about 9-in. depth, the supernatant was decanted off, and the biosolids were then mixed about twice a week by mechanical means until dry. Specifically, tractors equipped with screw conveyors in place of bulldozer blades scraped up the biosolids, mixed them, and diverted them to one side to form windrows. Extensive data on bacterial reductions were collected but not interpreted. Review of these data shows that the reduction in fecal coliform and salmonellae densities correlated much better with time on the bed than with other parameters, such as the ultimate solids content reached or the time of the year during which the material was dried. Drying on sand beds produced greater reductions in bacterial densities under the same environmental conditions than drying on mechanically mixed beds.

The sixth quarterly report of the Roswell research (Wilson and Co., 1989) gives information on temperature of the mechanically mixed beds versus soil temperature. Unfortunately, no details are given about the location of the temperature probes. The data show that the mechanically mixed beds were consistently 1-2°F cooler than the soil. Data were from the period October through January, so no information on the temperatures attained during the summer is available. This result indicates that in the prolonged drying of these mechanically mixed beds in relatively low windrows (about 1.5 ft high) temperature elevation caused by internal generation of heat or by absorption of solar radiation is not to be expected. The finding that bed temperatures are lower than soil temperatures is not unexpected since the biosolids beds will be cooled as water evaporates and, despite their dark surface, they will retain less solar radiation than soil. As the mechanically mixed beds dry, they become relatively fluffy and conduct heat much more poorly than compacted soil, so the radiant energy received does not penetrate deeply into the pile.

The Roswell experience showed that relatively rapid drying (e.g., six weeks) using low windrows (about 1.5 ft high) left high densities of salmonellae and fecal coliform even when the product reached 90% solids. As noted above, average temperatures during the windrow drying operation were actually lower than soil temperatures. When this information is considered, it appears that relatively extended drying periods of air drying are needed to adequately reduce bacteria and helminth ova unless some action (such as covering windrows with black polyethylene sheet) is used to improve heat absorption, reduce evaporation, and reduce convective heat loss.

Unless such actions are taken to raise windrow temperatures, the windrows will have to be dried over three months or longer during a favorable drying period. Destruction of helminth ova probably will have to occur by desiccation. Without these actions, there is less likelihood than in sand beds that temperatures in the windrows will rise to the point that helminth ova are rapidly destroyed.

B.8 Phillipp et al. (1985)

Phillipp et al. (1985) compared the effect of drying sludge on reed beds with the effect of drying on conventional sand beds, since there had been reports of reductions in bacterial densities in the vicinity of the rhizome of the reeds used in reed beds. The authors found no difference in the microbiological reductions produced by the two methods. Neither method was effective in reducing helminth ova densities. This paper is significant in that it discusses the only alternative low technology route, other than lagooning or air drying, to producing a pathogen-reduced biosolids product found in the European literature. Generally, interest in low-technology sludge disinfection systems is small in Europe because these systems tend to be land intensive, and there is little land available to site them.

B.9 Lucero-Ramirez and Malina (1998) and Lucero-Ramirez (2000)

Lucero-Ramirez and Malina (1998) and Lucero-Ramirez (2000) discussed reductions of indicator and pathogenic organisms throughout the solids treatment train at the Hornsby Bend Biosolids Management Plant (HBBMP) in Austin, Texas, over an approximate two-year period beginning in October 1996. The HBBMP processes raw solids from three WWTPs.

Raw solids entered an equalization tank and were then concentrated from 1.6% to 7.8% solids on a gravity belt thickener using a liquid polymer (Magnifloc 1598 C™) at a dose of 3.85 mL/kg dry solids. The thickened sludge was pumped to anaerobic digesters with no sludge heating (hydraulic retention time >75 days). The digested biosolids were air dried in open, paved, rolled concrete basins to about 20% solids. About 45% of the air-dried biosolids were applied to adjacent land where hay was grown as fodder. The remaining biosolids were composted aerobically.

The air drying basins were operated at a loading of approximately 200 tons/acre, dry weight basis, at an average initial applied liquid depth of 3 ft. The basin contents were turned three times a week using a Brown Bear™ or horizontal auger. The biosolids were held in the basins for 7-12 months, until they reached a solids content of about 20%.

Table B-1 presents pathogen-reduction results for the air drying system.

Table B-1. Pathogen and Pathogen Indicator Reduction at the Hornsby Bend Solids Management Plant, Austin, Texas.

Organism	Pathogen density		Log_{10} reduction
	Digested biosolids	Air-dried biosolids	
Total coliform[a] MPN/g DS[b]	27,300	545	1.69
Fecal coliform[a] MPN/g DS	12,900	45	2.45
Fecal streptococci[a] MPN/g DS	103,000	2,750	1.57
Salmonellae[a] MPN/4 g DS	4.08	0.64	0.79
Viable helminth ova, no./4 g DS			
Mean	7.7	2.2	0.54
Median	7.5	0.6	1.1
Range	0.8 to 15.0	0 to 13.1	--

[a] Geometric mean
[b] DS = dry solids

Geometric mean values for fecal coliform and salmonellae in air-dried biosolids were below the Class A standards. Fecal coliform values were below the Class A standard in 36 out of 41 samples, and salmonellae were rarely detected. When detected, they were below the Class A standard. The average density for viable helminth ova exceeded the Class A standard, while the median value did not. Viable helminth ova densities were below the Class A standard in 16 of 22 samples.

The investigators noted that air-drying operational practices were not uniform. Occasionally fresh biosolids were mixed with old biosolids. Therefore, the air-drying numbers presented should be considered preliminary. The authors felt that air drying of anaerobically digested biosolids under controlled conditions might be able to achieve Class A standards for both bacteria and helminth ova if mixing was avoided.

The report presented preliminary results on the effect of detention time on the density of viable helminth ova. After about 250 days of drying, ova densities were reduced from about 14 ova/4 g dry solids to <1 ova/4 g dry solids, the Class A limit. At the same time, solids content increased to about 15%. It is not clear whether these data were from full-scale experiments or experiments conducted in the laboratory. The investigators did not determine the dominant mechanism for pathogen kill (time, temperature, or desiccation). Desiccation as a mechanism for kill in this case seems unlikely, as the moisture content of the treated biosolids was much too high.

B.10 Ramey et al. (1996)

Ramey et al. (1996) reported that air-dried aerobically digested biosolids from the Louisville, Colorado, and the Pinery, Colorado, WWTPs successfully meet Class A criteria. Both plants are in the Denver area. The Louisville plant is a 114 l/sec (2.6 mgd) extended aeration plant with aerobic sludge digestion and centrifuge dewatering. The Pinery plant is a 22 l/sec (0.5 mgd) tertiary treatment plant with biological phosphorus removal, aerobic digestion, and belt filter press dewatering. Both plants air dry and store their dewatered biosolids to meet Class A requirements of the Part 503 regulations.

Louisville staff dewaters digested biosolids four to five days a week, and stockpiles the biosolids in several areas of the plant. A "Brown Bear" auger mixer machine divides the stockpiles into rows 150 ft long by 3.5 ft high. The rows are turned at least three times each summer and frequently more often. The city initially received complaints about odors when rows were being frequently turned. Complaints were reduced when the biosolids were stockpiled for at least six months prior to turning. Odor complaints also seemed to occur when high-nitrogen biosolids (about 4.5% total nitrogen) were being processed, but there were no complaints when the nitrogen content was low (0.8%). Louisville changed to storing biosolids in small piles 12-24 in. deep to assist air drying and reduce the nitrogen content.

Louisville's drying program produced good results in the reduction of pathogen and pathogen indicator organisms. Biosolids produced in the second quarter of 1994 were dried from 13% to 62% solids in three weeks during the summer. Fecal coliforms were reduced from 2.2 million MPN/g to 609 MPN/g and no salmonellae, helminth ova, or enteric viruses were detected. Winter drying (October through February) increased the solids content from about 16% to 50%, while reducing the volatile solids content from 58% to 35%. Salmonellae, fecal coliform, enteric virus, and helminth ova levels satisfied Class A criteria.

The Pinery staff typically operates its air drying process on a batch basis. The staff dewaters biosolids and transfers them to the air drying sites two to three times a week. The Pinery staff divides its piles into rows 200 ft long by 3.5 ft high. It has successfully used a modified air drying method of adding two loads of dried biosolids to one pile of recently dewatered cake to produce an initial pile solids content of 30-40%. This action has resulted in immediate heat generation and reduced odor significantly.

The Pinery reports good results in the reduction of fecal coliforms. For example, December to September drying increased one batch's solids content from 15-73%, while reducing the volatile solids content from 70% to 34%. Fecal coliforms dropped from 7.3 million MPN/g to negligible detection during that period. No information about other pathogens was provided.

Louisville and the Pinery appear to operate their systems as Alternative 4 processes, with each product batch analyzed for pathogens. The authors suggest that the air drying/storage process be considered by other plants in semi-arid locations (<15 in. rainfall per year). They cite processing costs of $75-150 per ton of dry solids.

B.11 Chicago Research Work, Tata et al. (2000); Lue-Hing et al. (1998)

Tata et al. (2000) reported on the results of three-year, full-scale studies at Chicago MWRD's Calumet and Stickney wastewater reclamation plants. Each plant uses two similar but distinct SPTs. Anaerobic digestion (35°C [95°F], 20 days SRT) is the first step in the SPT. Anaerobically digested biosolids are further processed by either of two methods:

- *High-solids SPT (HSSPT).* Anaerobically digested biosolids are dewatered to a 25% solids cake with centrifuges. The cake is then transported by rail to lagoons where it is stored and aged to further inactivate pathogens for at least 18 months (18 months of minimum unfed storage time in lagoons). After storage and stabilization in the lagoons, the cake is removed from the lagoons and air dried to approximately 60% solids on paved beds.
- *Low-solids SPT (LSSPT).* Anaerobically digested biosolids are transferred to lagoons which are filled in cycles over one to three years. Batches are added layer-by-layer until the biosolids fill the lagoon, at which point the biosolids have concentrated to about 15% solids. After storage and stabilization in the lagoons (at least 18 months of unfed storage time), the

biosolids are removed from the lagoons and air-dried to approximately 60% solids on paved beds.

Previously MRWD had operated the SPTs to produce a 60% solids air-dried product for beneficial use. Pathogen reduction had not been the primary goal. In the study reported, the district optimized and codified SPT protocols and imposed additional controls to ensure the modified SPTs would meet Class A criteria.

This appendix discusses the district's air drying operations. The District's lagoon operations are discussed in Appendix A and the overall system operations in Appendix D.

Air drying took place from April through November. Any batch from the HSSPT and LSSPT lagoons applied to the drying areas was dried to a minimum solids concentration of 60% without any further additions to the drying beds. The biosolids from the HSSPT and LSSPT lagoons were air dried at area loading rates of no more than 92 kg dry solids/sq m and 52 kg dry solids/sq m (410 dry tons/acre and 230 dry tons/acre), respectively. Actual loadings for the Calumet and Stickney HSSPT beds were 59.9 kg/sq m and 96.2 kg/sq m (267 tons/acre and 429 tons/acre), respectively. Loadings for the Calumet and Stickney LSSPT beds were 59.9 kg/sq m and 41.9 kg/sq m (287 tons/acre and 187 tons/acre), respectively. Biosolids depths were no more than 460 mm and 380 mm (18 in. and 15 in.), respectively, to be consistent with the above loadings. The drying beds were turned, aerated, and agitated an average of three times a week using tractor equipment with a horizontal auger or a tiller. When the solids concentration reached about 35%, the biosolids were formed into windrows to allow free drainage of rainfall and to encourage development of higher in-bed biosolids temperatures. Drying times for LSSPT biosolids averaged 63 days, for HSSPT biosolids 67 days.

Digester feed, digester draw, lagoon draw and air-dried product were analyzed for fecal coliform, salmonellae, viruses, and viable helminth ova. Table B-2 presents results for reduction of these organisms in the air drying process.

All codified LSSPT and HSSPT air-dried samples achieved Class A standards for pathogens and indicator organisms. Lagoon draw and air dried final product virus and helminth ova densities were very low, meeting Class A requirements. Therefore it was difficult to directly demonstrate that air drying produced high log reductions of these organisms.

Table B-2. Reduction of Pathogens and Pathogen Indicators in Calumet and Stickney Air Drying Systems.

System	Organism	Pathogen Density: Lagoon Draw (codified)[a]	Pathogen Density: Air-dried Final Product (codified)	Log_{10} Reduction
Calumet HSSTP	Fecal coliforms, MPN/g DS[b,c]	105	5.58	1.27
	Salmonella sp, MPN/4 g DS[d]	0.828	0.419	0.30
	Virus, PFU/4 g DS[d]	0.00	0.0	--
	Viable helminth ova, no./4 g DS[d]	7.44×10^{-3}	1.54×10^{-3}	0.68
Calumet LSSTP	Fecal coliforms, MPN/g DS	68.5	17.1	0.60
	Salmonella sp, MPN/4 g DS	2.9	0.673	0.63
	Virus, PFU/4 g DS	0.00	0.0	--
	Viable helminth ova, no./4 g DS	0.0625	3.75×10^{-3}	1.22
Stickney HSSPT	Fecal coliforms, MPN/g DS	188	18.3	1.01
	Salmonella sp, MPN/4 g DS	0.276	0.506	(0.26)
	Virus, PFU/4 g DS	0.00	0.0	--
	Viable helminth ova, no./4 g DS	0.784	0.0191	1.61
Stickney LSSPT	Fecal coliforms, MPN/g DS	151	25.6	0.77
	Salmonella sp, MPN/4 g DS	0.533	0.532	0.00
	Virus, PFU/4 g DS	0.00	0.0	--
	Viable helminth ova, no./4 g DS	0.0394	0.0132	0.48

a Codified means operating at conditions as stringent or more stringent than specified by the operating protocols.
b Geometric mean
c DS = dry solids
d Arithmetic mean
Source: Tata et al. 2000

Bergman and Rohloff (1996) reported costs for Chicago's sludge-treatment operations in the years 1992 through 1995, including air drying. Table B-3 summarizes their information.

Table B-3. Average costs for air drying operations at the Metropolitan Water Reclamation District of Greater Chicago (1992 through 1995).

Operation	Average cost, $/ton DS[a]: LSSPT	Average cost, $/ton DS[a]: HSSPT
Transport to drying	21.39	31.97
Air drying	28.57	32.73
Loading for transport to disposal	3.09	3.24
Transport to disposal	14.75	5.14

a $/tonne DS = 1.10 x $/ton DS
DS= dry soilids

B.12 McHaney et al. (1994)

McHaney et al. (1994) reported on results from the solids stabilization section of the San Jose/Santa Clara, California, Water Pollution Control Plant. The stabilization section combines anaerobic digestion, lagoon storage, and air drying processes. This appendix reports on air drying operations. Appendix A describes lagoon storage operations and Appendix D the combined lagoon storage/air drying process.

The drying beds are filled to a depth of 46-51 cm (18-20 in.) with dredged lagoon slurry biosolids. For the first two or three weeks continued settling occurs and sun and wind cause the overlying liquid to evaporate. Then a crust forms over the biosolids, preventing further evaporation. Then rubber-tired tractors equipped with mixing augers turn the biosolids, breaking up the crust, allowing evaporation to continue. The biosolids concentration reaches 60-75% after approximately 12 weeks of uninterrupted drying. The biosolids are typically mixed with the augurs three to four times during this period. The dried biosolids are moved by truck to a stockpile area where they reside for substantial periods. Subsequently, they are transported off site for beneficial use.

Preliminary analyses from March 1994 showed that air drying reduced fecal coliform from 129 MPN/ g dry solids to <2 MPN/ g dry solids and salmonellae from 159 MPN/ 4 g dry solids to <2 MPN/ 4g dry solids. The enterovirus density in the air-dried product was 1.5 PFU/ 4g dry solids. Viable helminth ova were not detected.

The above analyses were completed before plant staff adopted a period of unfed storage time in its lagoons (in the mid 1990s). Also, plant staff modified air drying operations in the mid 1990s, so fecal coliform and salmonellae results are likely to differ from those noted above. For combined lagoon/air-drying system operation and performance at San Jose since the mid 1990s, see Garner (2004) in Appendix D.

B.13 Miami-Dade, Florida (1998)

In 1997, the Miami-Dade Water and Sewer Department (1998) undertook pathogen testing at its South Plant using small-scale air drying piles starting with anaerobically digested, dewatered biosolids at about 24% solids content. Several different arrangements were tested, including immediate mechanically-mixed air drying, as well as covering the piles with plastic for two to three months prior to air drying. Temperatures within the piles were monitored and showed that under many conditions, temperatures were elevated above ambient temperatures which would assist in more rapid pathogen destruction. Air-dried biosolids were typically in the range of 60-80% solids after several or many weeks of drying. Fecal coliform data, however, were scattered and inconsistent, and fecal coliform regrowth appeared to occur at times.

APPENDIX C

REVIEW OF THE CAKE STORAGE LITERATURE

The literature the project team reviewed concerning pathogen reduction during cake storage of municipal biosolids is summarized below.

C.1 Ahmed and Sorensen (1997) and Ahmed and Sorensen (1995)

Ahmed and Sorensen (1997) investigated the fate of microorganisms in anaerobically and aerobically digested biosolids that had been dewatered and stored for long periods in windrows 1.2 m (4 ft) to 1.5 m (5 ft) high at five WWTPs in Utah. Two to six (usually four) windrows were set up at each location. One set of windrows was turned once a month (in some cases twice a month), and the other set was left unturned. In each set of windrows, one was aligned north–south and the other east–west to test the effects of prevailing winds and the sun. To follow the decline of microorganism densities, cake was spiked with microorganisms and the cake was charged into 1.9-cm (0.75 in) diameter cylinders, 0.91 m (3 ft) long, constructed of fiberglass door screening. The microorganisms were *Salmonella typhimunium, Yersinia enterocolitica, Campylobacter jejuni,* bacteriophage f-2, and *A. suum* ova. The cylinders were introduced into the windrows vertically at uniform spacing along the apex of the windrows. The porous cylinders allowed moisture and heat transport between the bulk of the windrow and the spiked biosolids but not solids transport. Periodically, one of the cylinders was removed for analysis.

The experiments began in the spring. All of the windrows steadily rose in temperature until about the end of August. All bacterial pathogens were destroyed, generally within 30-70 days and, at most, 100 days. Fecal coliform, which was determined on samples collected from the bulk of the windrows and not from the sample tubes, fell to insignificant densities (0.3 MPN/g dry solids) within 96 days. The unturned and turned windrows showed about the same behavior. No information was obtained on the decline of bacteriophage f-2 because of experimental difficulties.

The temperatures of the unturned piles closely followed the average daily temperatures, averaging between 25 °C and 32°C (77 °F and 90°F) after 90 days for the five plants. At two plants, the turned windrows rose to considerably higher temperatures than the unturned windrows, in one case reaching 40°C higher (maximum temperature 65°C [149°F]) and at the other 15°C higher (maximum temperature 44°C [111°F]). Initially, helminth ova densities fell more rapidly in the higher-temperature piles than the lower-temperature piles. However, this advantage was not maintained and it took 300-400 days of storage to reduce helminth ova to nondetectable levels regardless of temperatures achieved in the piles. The investigators reported total intact helminth ova densities, not viable helminth ova densities. Since the latter should be less than the former, it is likely that the final product satisfied Class A limits for viable helminth ova. Table C-1 summarizes cake storage results for the five treatment plants.

Table C-1. Summary of Pathogen Kill by Cake Storage in Five Utah Plants.

Plant	Prior Treatment	Pathogen	Organism Density, Log 10 no./g DS Initial	Final	Log10 Reductions	Number of Turnings per Month: 0	1	2
						Time to Non-detect, Days		
Central Weber	AN[a], VF[b]	FC[c]	2 to 3	0.5	2 to 3	100	60	100
		S[d]	5 to 7	0.5	5 to 7	35	70	70
		TIHO[e]	3.0 to 4.5	0.2	2.8 to 4.3	330	330	330
South Davis	AN, DB[f]	FC	NDT[g]	0.5	NDT	--	--	--
		S	5.5	0.5	5.5	40	40	ND[h]
		TIHO	2.6	0.2	2.4	400	325	ND
Central Davis	AE[i], DB	FC	ND	ND	ND	ND	ND	ND
		S	5.5	0.5.	5.5	25	55	ND
		TIHO	2.9	0.2	2.7	330	330	ND
Timpanogos	AE, DB	FC	3.2	0.5	3.2	70	90	ND
		S	6	0.5	6	90	90	ND
		TIHO	2.3	0.2 to 0.8	1.5 to 2.3	ND	325	ND
Snyderville	AE, BP[j]	FC	ND	ND	ND	ND	ND	ND
		S	4	0.5	4	25	25	ND
		TIHO	2.2	0.2	2.0	300	300	ND

a AN = anaerobic digestion
b VF = vacuum filtration
c FC = fecal coliform
d S = salmonellae
e TIHO = total intact helminth ova
f DB = drying beds
g NDT = none detected
h ND = no data
i AE = aerobic digestion
j BP = belt press

The authors derived the following relationship to predict the peak biosolids temperature:

$$T_p = -30.44 + 8.3\ (C/N) + 4.8\ MIX \qquad (C.1)$$

where: T_p = peak temperature, °C
C/N = carbon to nitrogen weight ratio
MIX = number of turnings per month

Other factors (pile moisture content, pile orientation, and VSS concentration) did not significantly affect peak pile temperatures.

Objections could be raised to the conclusions of this work because, except for fecal coliform, microbial measurements were only made in the center of the windrows. The exterior of the windrows could have experienced much different conditions. Examination of the report shows substantial variations in temperature across the windrows. Another objection is that the *Ascaris* ova used in the experiment were extracted from gravid worms and may not have had the fully developed resistant coating that would be found on ova collected from fecal matter. On the other hand, it should be noted that the unturned windrows produced the same microbiological reduction as the turned windrows, although they were a few degrees cooler, so minor temperature differences within a windrow might not have much influence on results. Also, if the elevated temperature was a significant contributing factor to the destruction of the ova, the lack of a resistant coat may not be a relevant objection since this thin coat would not protect the ova from a temperature increase.

Ahmed and Sorensen (1995) reported the results of a bench-scale investigation of the kinetics of decay in densities of the same organisms noted above as well as poliovirus spiked into biosolids and stored under aerobic and anaerobic conditions at four temperatures (5°C, 22°C, 38°C, and 50°C [41°F, 72°F, 100°F, and 122°F]). Three storage runs were made using different batches of anaerobically digested biosolids that had previously been vacuum filtered and air dried for three to five months. In the process of spiking, water was added to the biosolids. For Runs 1, 2 and 3, moisture content after spiking at the start of the tests were 67.3%, 47.7%, and 50.0%, respectively. Moisture contents varied by <5% throughout the experiments. The extractable ammonium for Runs 1, 2, and 3 was 0.5%, 0.25%, and 1.25%, respectively.

The study identified data relevant to pathogen declines under environmental conditions usually encountered during optimum drying periods. At 22°C (72°F), *C. jejuni* declined to below detection limits (0.6 MPN/ g dry solids) in 10 days. At this temperature, polioviruses and bacteriophage f2 declined more rapidly (over 4 logs in 30 days) than the other organisms, except for *C. jejuni.* (Bacteriophage f2 is hardier than enteroviruses under environmental conditions and thus a good indicator organism for enteroviruses.) *S. typhimurium* and *Y. enterocolitica* declined to below detection limits (0.6 MPN/ g dry solids) in about 30 days except for Run 1, in which the biosolids had a higher moisture content (67%). In this case, it took about 50 days to reduce densities by more than 4 logs. All densities declined faster at higher temperatures. Total *Ascaris* ova were extremely hardy at 22°C (72°F), declining less than a log after 60 days. (Again, the authors reported the density of total *Ascaris* ova, not viable *Ascaris* ova.)

The data showed no substantial differences between kills at aerobic and anaerobic conditions, but there were differences in results that could be attributed to moisture content or ammonium content. The effect of moisture for *S. typhimurium* and *Y. enterocolitica* has been noted above. Another example is the *Ascaris* results at 50°C (122°F). Results for Runs 2 and 3 were divergent, showing a reduction of ≤1 log after five days for both anaerobic and aerobic conditions of Run 2 and over 2 logs after five days for Run 3. The biosolids for both runs had about the same moisture content (about 50%), total nitrogen was about the same (3.6%), but the initial ammonium content was 0.25% for the Run 2 biosolids and 1.32% for the Run 3 biosolids, which showed the greater log reductions. The effect of ammonium content was expected, since ammonia is a known ovicide. The effect was not noticeable at lower temperatures, perhaps because the experiment terminated after only 60 days. The data show examples of how higher ammonia and lower moisture concentrations accelerated kill of all microorganisms except *S. typhimurium*. Poliovirus, for example, declined far faster in every case for Run 3, the high ammonium biosolids, than for Run 2.

The authors pooled their data for the different microorganisms, based on the linearity of the correlation between log microbial density and time, to determine the effect of temperature on the kinetic constant, K. K. the number of $\log_{10}$ pathogen reductions that occur in one day.

First, the authors simulated pathogen density versus time data for each organism at each temperature tested using the following equation:

$$\log_{10} (N/N_0) = -Kt + b \qquad \text{(C.2)}$$

where: N = organism concentration at time t

N_0 = initial organism concentration (at time zero)

K = temperature-dependent kinetic constant (slope of $\log_{10}$ (N/N_0) vs. t plot)
t = time, days
b = intercept of $\log_{10}$ (N/N_0) axis at time zero

Then they used the Arrhenius relationship to estimate the influence of temperature on each organism's kinetic constant:

$$\ln K = \ln C - E_d/RT_k \qquad \text{(C.3)}$$

where: C = a constant
E_d = Arrhenius activation energy
R = universal gas constant
T_k = absolute temperature, degrees Kelvin

The report provides estimates of K for the organisms and temperature range tested. The authors apparently did not consider effects of moisture and extractable ammonium content. Consequently, their kinetic data should be used with caution.

C.2 Gibbs et al. (1995)

Gibbs et al. (1995) tested storage as a possible low-cost method of pathogen reduction prior to biosolids reuse in Australia. They created a 40 m^3 pile of anaerobically digested (mesophilic digestion), mechanically dewatered biosolids. The initial pile was about 1 m high on a sandy soil above a limestone base. The initial solids concentration was about 12%, increasing to about 30% after 53 weeks of storage. The pile was not covered.

They conducted about 10 samplings over the 53-week test period, with earlier samples more closely spaced together. Initially, five replicate samples were collected from about 20-40 cm (7.8-15.7 in.) and 40-60 cm (15.7-23.6 in.) below the biosolids surface. Sampling at the 40-60 cm level was discontinued after 24 weeks, as pile shrinkage eliminated that sampling point. The authors analyzed the biosolids for fecal coliforms, fecal streptococci, *Giardia*, enteroviruses, salmonellae, total suspended solids (TSS), and VSS, and pH.

Because the authors did not detect enteroviruses in any of the first four samplings, no virus measurements were made in subsequent samplings. Lack of viruses was unexpected, since they had been detected in biosolids samples in previous studies.

Salmonellae and fecal coliforms were not detected in biosolids sampled after 34 weeks of storage (salmonellae were not detected after 12 weeks of storage). Subsequently, both regrew to densities that exceeded in the initial charge. Calculations indicate regrown densities of both organisms exceeded Class A limits by wide margins. Fecal streptococci densities were only marginally reduced during the first 25 weeks of storage; subsequently, they regrew to densities that exceeded the initial charge. The authors speculated that regrowth came either from low numbers of non-detected organisms or from re-inoculation by animal feces or insect contamination. However, rainfall after a long, dry period and just before sampling may have contributed to regrowth or washed feces or bacteria into the sampling areas.

Giardia densities were not reduced significantly. However, the authors noted that the monoclonal test used to determine *Giardia* densities does not distinguish between dead and living cysts. It is possible that cysts were intact but not viable. Nevertheless, the authors were concerned that survival of *Giardia* is a potential hazard for biosolids disposal and that there should be further research to determine extent of survival in land application of biosolids.

The authors noted that biosolids temperatures did not exceed 23°C (73°F), moisture content did not drop below 70%, and VSS concentrations remained high (more than 70% of TSS). The low temperatures and high moisture contents did not favor high pathogen kill. The high VSS and moisture contents facilitated pathogen regrowth. Therefore, conditions did not favor production of low-pathogen biosolids.

C.3 Al-Ghazali and Al-Azawi (1988)

Al-Ghazali and Al-Azawi (1988) measured the decline of the density of the bacteria *Listeria monocytogenes* in a large heap of digested biosolids that had previously been spread on drying beds for two weeks. The work was conducted in Iraq, which has a hot, arid climate. The authors took samples and temperature measurements at the surface of the heap and inside the heap (50 cm [20 in.] deep).

Initial samples were moist (35% surface moisture and 43% moisture inside the heap) and contained 4 counts/g solids at the surface and 9 counts/g solids inside the heap of *L. monocytogenes*. The density of *L. monocytogenes* at the surface declined to zero within six weeks. During this period, the surface moisture content dropped and rose again, from 35% to 2% to 8%, and the surface temperature rose from 28°C (82°F) to 32°C (90°F). The surface pathogen density remained non-detectable for the rest of the 23-week experiment. The surface temperature continued to rise, reaching 42°C (108°F) during week 23.

The *L. monocytogenes* density on the inside of the heap fell to zero during week 8. By that time, the temperature of the inside sample had risen from 48°C (118°F) to 54°C (129°F), and the moisture content had fallen from 43% to 11%. The pathogen density remained at non-detectable for the remainder of the experiment. The moisture content and temperature also declined to 9% and 45°C (113°F), respectively.

The most noteworthy aspect of this research was the heat tolerance of *L. monocytogenes*. It was detected after seven weeks storage at high temperatures (43-56°C; 109-133°F). Further, it dropped to undetectable concentrations when the moisture content dropped below about 11%.

C.4 Jepsen et al. (1997)

Jepsen et al. (1997) summarized the results of several treatment processes (including storage) on fecal streptococci and salmonellae in Denmark. In the storage experiments, the effects of time were studied by storing 5 m^3 (6.5 yd^3) of anaerobically digested biosolids and 5 m^3 (6.5 yd^3) of aerobically digested biosolids in separate containers. Fecal streptococci and salmonellae populations were tracked for six months. Samples were taken as vertical profiles, excluding the top 20 cm, and then mixed prior to analyses. The investigators analyzed the effects of both warm- (biosolids temperature 20-30°C [68-86°F]) and cold-weather storage (biosolids temperatures 0-10°C [32-50°F]). The solids content of the anaerobic biosolids was 28%. The solids content of the aerobic biosolids was 20%.

Warm-weather storage effects were as follows.

- Fecal streptococci in aerobically digested biosolids reduced from 10^6 to 10^4/g dry solids in 2 months, but there was no further reduction at month 3.
- Fecal streptococci in anaerobically digested biosolids reduced from 20,000/g to 20/g in 4 months.
- Salmonellae in aerobically digested biosolids reduced from 800,000 MPN/100 g to below the detection limit (1 MPN/100 g) in 1 month.
- Salmonellae in anaerobically digested biosolids reduced from 300 MPN/100 g to below the detection limit in one month.

Cold-weather storage results were as follows:

- Fecal streptococci in aerobically digested biosolids reduced by 1 log (from 10^6 to 10^5/g dry solids) in over six months of storage.
- Fecal streptococci in anaerobically digested biosolids reduced from 200,000/g to 2000/g in about five months, but there appeared to be regrowth in month 5.
- Salmonellae in aerobically digested biosolids reduced from 10,000 MPN/100 g to about 30 MPN/100 g in six months.
- Salmonellae in anaerobically digested biosolids reduced from 10,000 MPN/100 g to below the detection limit in about four months.

The authors concluded that storage at temperatures exceeding 20°C (68°F) could greatly improve the hygienic properties of biosolids, but temperatures less than10°C (50°F) had only minor effects, even with long periods of storage.

APPENDIX D

REVIEW OF THE LITERATURE ON COMBINED LAGOON STORAGE/AIR DRYING SYSTEMS

The literature the project team reviewed concerning pathogen reduction in combined lagoon storage and air drying systems treating municipal biosolids is summarized below.

D.1 Tata et al. (2000), Lue-Hing et al. (1998), and Bergman and Rohloff (1996)

Tata et al. (2000) reported on the results of three-year, full-scale studies at MWRD's Calumet and Stickney wastewater reclamation plants. Each plant uses two similar but distinct SPTs. Anaerobic digestion (35°C [95°F], 20 days SRT) is the first step in the SPT. Anaerobically digested biosolids are further processed by either of two methods:

- *High-solids SPT (HSSPT).* Anaerobically digested biosolids are dewatered to a 25% cake with centrifuges. The cake is then transported by rail to lagoons where it is stored and aged to further inactivate pathogens for at least 18 months (18 months of minimum unfed storage time in lagoons). After storage and stabilization in the lagoons, the cake is removed from the lagoons and air dried to approximately 60% solids on paved beds.
- *Low-solids SPT (LSSPT).* Anaerobically digested biosolids are transferred to lagoons which are filled in cycles over one to three years. Batches are added layer-by-layer until the biosolids fill the lagoon, at which point the biosolids have concentrated to 15% solids. After storage and stabilization in the lagoons (at least 18 months of unfed storage time), the biosolids are removed from the lagoons and air dried to approximately 60% solids on paved beds.

Previously, MRWD had operated the SPTs to produce a 60% solids air-dried product for beneficial use. Pathogen reduction had not been the primary goal. In the study, the district optimized and codified SPT protocols and imposed additional controls to ensure the modified SPTs would meet Class A criteria.

Minimum holding times for HSSPT and LSSPT lagoons were set at 1.5 years after last placement of biosolids. Actual holding times were longer. The average holding times for Calumet and Stickney LSSPT lagoons were 2.5 years and 2.0 years, respectively. The average detention times for Calumet and Stickney HSSPT lagoons were 1.6 years. Measured sludge layer temperatures are shown in Table D-1.

Table D-1. Sludge Layer Temperatures (°C) in the Calumet and Stickney Lagoons.

	Calumet		Stickney	
Item	LSSPT	HSSPT	LSSPT	HSSPT
Low temperature	12	7	6	6
High temperature	24	21	21	20
Mean temperature	15	13	13	14

Source: Lue-Hing et al. 1998

Air drying took place from April through November. Any batch from the HSSPT and LSSPT lagoons applied to the drying areas was dried to a minimum solids content of 60% without any further additions to the drying beds. Biosolids from the HSSPT and LSSPT lagoons were air dried at area loadings of no more than 92 kg dry solids/sq m and 52 kg dry solids/sq m (410 dry tons/acre and 230 dry tons/acre), respectively. Actual loadings for the Calumet and Stickney HSSPT beds were 59.9 kg/sq m and 96.2 kg/sq m (267 tons/acre and 429 tons/acre), respectively. Loadings for the Calumet and Stickney LSSPT beds were 59.9 kg/sq m and 41.9 kg/sq m (287 tons/acre and 187 tons/acre), respectively. Biosolids depths were no more than 460 mm and 380 mm (18 in. and 15 in.), respectively, to be consistent with the above loadings. The drying beds were turned, aerated, and agitated an average of three times a week using tractors with horizontal augers or tillers. When the solids concentration reached about 35%, the biosolids were formed into windrows to allow free drainage of rainfall and to encourage development of higher in-bed temperatures. Drying times averaged 63 days for LSSPT biosolids and 67 days for HSSPT biosolids.

Digester feed, digester draw, lagoon draw, and air-dried product were analyzed for fecal coliform, salmonellae, viruses, and viable helminth ova. Table D-2 presents results for removal of these organisms across the combined lagoon/air drying system. (Removals across the lagoons and air drying processes individually are presented in Appendices A and B, respectively.)

All codified LSSPT and HSSPT air-dried samples achieved Class A standards for pathogens and indicator organisms. Lagoon draw and air dried final product virus and helminth ova densities were very low, within Class A requirements. Therefore, it was difficult to directly demonstrate that lagoon storage produced high log reductions of these organisms.

Table D-2. Reduction of Pathogens and Pathogen Indicators in Calumet and Stickney Combined Lagoon/Air Drying System.

System	Organism	Pathogen Density Digester Draw	 Air-dried Final Product (Codified)[a]	Log_{10} Reduction
Calumet HSSTP	Fecal coliforms, MPN/g DS[b,c]	6.78×10^4	5.58	4.08
	Salmonella sp, MPN/4 g DS[d]	74.6	0.419	1.87
	Virus, PFU/4 g DS[d]	0.428	0.0	--
	Viable helminth ova, no./4 g DS[d]	0.286	1.54×10^{-3}	2.26
Calumet LSSTP	Fecal coliforms, MPN/g DS	6.78×10^4	17.1	3.59
	Salmonella sp, MPN/4 g DS	74.6	0.673	2.05
	Virus, PFU/4 g DS	0.428	0.0	--
	Viable helminth ova, no./4 g DS	0.286	3.75×10^{-3}	1.88
Stickney HSSPT	Fecal coliforms, MPN/g DS	2.02×10^5	18.3	4.04
	Salmonella sp, MPN/4 g DS	25.7	0.506	1.70
	Virus, PFU/4 g DS	0.121	0.0	--
	Viable helminth ova, no./4 g DS	0.379	0.0191	1.30
Stickney LSSPT	Fecal coliforms, MPN/g DS	2.02×10^5	25.6	3.90
	Salmonella sp, MPN/4 g DS	25.7	0.532	1.68
	Virus, PFU/4 g DS	0.121	0.0	--
	Viable helminth ova, no./4 g DS	0.379	0.0132	1.46

a Codified means operating at conditions as stringent or more stringent than specified by the operating protocols.
b Geometric mean
c DS = dry solids
d Arithmetic mean
Source: Tata et al. 2000

Bergman and Rohloff (1996) reported costs for Chicago's biosolids-treatment operations in the years 1992 through 1995, including costs for lagooning and air drying. Table D-3 summarizes their information.

Table D-3. Average Costs for Combined Lagooning/Air Drying Processes at the Metropolitan Water Reclamation District of Greater Chicago (1992-1995).

	Average Cost, $/ton DS[a]	
Operation	**LSSPT**	**HSSPT**
Lagoon dewatering	11.08	0.44
Lagoon outloading	14.09	10.00
Transport to drying	21.39	31.97
Air drying	28.57	32.73
Loading for transport to disposal	3.09	3.24
Transport to disposal	14.75	5.14
Total	92.99	85.83

[a] $/tonne DS = 1.10 x $/ton DS

D.2 McHaney et al. (1994) and Garner (2004)

McHaney et al. (1994) reported on results from the biosolids stabilization system of the San Jose/Santa Clara, California, Water Pollution Control Plant. The stabilization system combined anaerobic digestion, lagoon storage, and air drying processes. Raw sludge and thickened waste sludge from the suspended growth secondary and nitrification units were anaerobically digested at 35°C (95°F) for up to 40 days. The digested biosolids (about 1.5% solids) were pumped to almost 30 storage lagoons (3.1 m [10 ft] deep) of various sizes (several acres each). Lagoon supernatant was returned to the plant headworks. Lagoon water temperatures varied from 15°C to 18°C (59 °F to 64°F). Biosolids were harvested by floating dredges, which pumped the solids to air/solar drying beds. The dredged biosolids were about 3-6% solids.

The drying beds were filled to a depth of 46-51 cm (18-20 in.). For the first two or three weeks continued settling occurred and sun and wind caused the overlying liquid to evaporate. Then a crust formed over the biosolids, preventing further evaporation. Then rubber-tired tractors equipped with mixing augers turned the biosolids, breaking up the crust, allowing evaporation to continue. The dried biosolids were moved by truck to a stockpile area where each year's production was segregated into piles. Subsequently, the biosolids were transported off-site for beneficial use. The annual cost of lagoon stabilization and solar drying averaged about $10/yd^3. If one assumes a product solids content of 60% and product density of 692 kg/m^3 (45 lb/ft^3), the cost was about $31/tonne ($28/ton) dry solids.

Table D-4 shows results of biosolids analyses from March 1994.

Table D-4. Preliminary Results of Biosolids Analyses at the San Jose/Santa Clara Water Pollution Control Plant, March 1994.

Position	% solids	% VSS	Fecal Coliforms, MPN/g DS	Salmonellae, MPN/4 g DS	Enterovirus, PFU/4 g DS	Helminth ova/g DS
Predigestion	3.82	78.8	2.7×10^7	402	5	NR[a]
Digester effluent	1.75	63.7	3.9×10^6	167	2	NR
Lagoon-dredged biosolids	3.17	51.4	129	159	NR	0
Air-dried biosolids	61.45	12.2[b]	2	<2	1.5	0

[a] NR = not reported

[b] Low VSS content is surmised to be due to entrainment of drying bed soils into the dried biosolids

Source: McHaney et al., 1994

Note that lagoon storage dramatically diminished the density of fecal coliforms but not the density of salmonellae. The lagoon-dredged biosolids satisfied the Class A criterion for fecal coliform but not the Class A criterion for samonellae.

In the mid-1990s, plant staff changed lagoon operations to include a two-year period of unfed storage (Garner, 2004). This was a direct attempt to upgrade the final product to meet Class A biosolids standards. Each of several lagoon groupings since that time have been fed/decanted for one year, remained unfed for two years, and then dredged in the fourth year. The two-year unfed lagoon storage time was selected using a conservative approach after evaluating research available in the mid-1990s. No work has been done since then to possibly reduce or fine tune the two-year unfed storage time since the plant has sufficient numbers of lagoons to accommodate this operational strategy.

Air-drying system changes in the mid-1990s included filling the drying beds to deeper levels initially and using different mixing/turning equipment that does not entrain as much bottom soil as previously. In recent years, drying-bed biosolids are typically turned several or many times in the three- to five-month drying cycle. The solids content reaches 70-75% during this period in the warm, dry San Jose summers.

After implementing the two-year period of unfed lagoon storage in the mid-1990s, the plant has produced Class A product every year (Class A status is verified annually by pathogen testing of each year's air-dried product pile). This large plant produces about 20,000 tons/year (dry weight basis) of final biosolids product.

APPENDIX E

LITERATURE REVIEW-EFFECT OF AMMONIA ON PATHOGEN KILL

The literature the project team reviewed concerning the effects of ammonia on the kill of pathogens and indicator bacteria in municipal biosolids is summarized below.

E.1 Ward and Ashley (1977)

Ward and Ashley (1977) identified the un-ionized form of ammonia (NH_3) as the sole agent responsible for inactivation of type 1 CHAT poliovirus in mesophilically digested municipal biosolids. They first isolated the agent from the digested biosolids by dialysis into de-ionized water. They found the virucidal activity disappeared when they tried to concentrate the agent by heating the de-ionized water to 50°C (122°F) under a stream of nitrogen, indicating the agent was volatile.

Then, they distilled some digested biosolids. More than 95% of the virucidal activity was concentrated in the first 15% of the distillate. The distillate was contacted with activated charcoal, which removed a great deal of its pungent, disagreeable odor, but not its virucidal activity. The charcoal-treated distillate retained the clear, unmistakable odor of ammonia.

The researchers next determined virucidal activity of a buffered ammonium chloride solution over the pH range 7 to 10. They compared its activity against activities of control solutions that contained only buffer or only NaCl over the same pH range to eliminate the possibility that the virucidal activity was caused by the buffer or salts. Virus reductions in the buffered ammonium chloride solution were 0.04 logs, 0.34 logs, 3.9 logs, and >5.8 logs at pH of 7, 8, 9, and 10, respectively. They estimated the un-ionized fraction of ammonia was <1%, 5%, 41%, and 87% of total ammonia at pH of 7, 8, 9, and 10, respectively. Virus kills in the control solutions were negligible compared to the kills in the ammonium chloride solution. Furthermore, the distillate from the previous experiment (which had the same ammonia content as the buffered ammonium chloride solution) had almost identical virus reductions at the same pH values. These results identified un-ionized ammonia as the effective virucide.

The investigators also showed that other strains of poliovirus (poliovirus type 1 strain Mahoney, poliovirus type 2 strain 712, cocksackie A13, coxsackie B1, and echovirus 11) are also readily inactivated by ammonia. These viruses all belong to the pivornavirus group. However, reovirus type 3, a virus belonging to another group, was quite resistant.

E.2 Reimers et al. (1986)

Reimers et al. (1986) evaluated a number of techniques for disinfecting biosolids, such as ultrasonic treatment, high energy irradiation, and chemical treatment. Technology was mostly sophisticated or expensive. One option that could be classified as low technology was treatment with gaseous ammonia (fertilization of soil with gaseous ammonia is practiced on many farms) or ammonium sulfate. About 0.1% ammonium sulfate (as ammonia), based on solids content, destroyed *Ascaris* ova. About 1% gaseous ammonia was needed to inactivate the ova.

E.3 Ghiglietti et al. (1995)

Ghigliette et al. (1995) evaluated the effect of ammonium hydroxide, time, and temperature on the survival of helminth ova in a laboratory experiment. The investigators obtained *A. suum* eggs (from a pig), *A. lumricoides* eggs (from a worm taken from an infected person and also worms from human feces) and *Trichuris muris* eggs (from feces of infected mice). The eggs were suspended in 30% NH_4OH (pH 11.9) for periods up to 21 days, at temperatures of 22°C, 30°C, and 40°C (72°F, 86°F, and 104°F). Control samples were maintained at the same temperatures without ammonia (eggs were stored in pH 7 saline solution).

At 22°C (72°F), ammonia and time had no effect on the viability of *A. suum* and *A. lumbricoides* eggs. Ammonia reduced the viability of *T. muris* eggs to 82%, 30%, and 3% after storage for 7, 14, and 21 days, respectively. The control eggs remained viable for the duration of the test.

At 30°C (86°F), the viability of *A. suum, A. lumbricoides,* and *T. muris* eggs dropped to 98%, 93%, and 0%, respectively after seven days of storage and to 2%, 1%, and 0% after 14 days of storage. The control eggs remained viable for the duration of the test.

At 40°C (104°F), viability of *A. suum* and *A. lumbricoides* eggs were 0% after 7 days of storage. The viability of *A. suum* and *A. lumbricoides* control eggs were 19% and 10%, respectively, after seven days of storage and 0% after 14 days of storage. Therefore, temperature alone was sufficient to partially inactivate these eggs at storage times of 14 days. It was not possible to evaluate the viability of *T. muris* eggs at 40°C because this temperature is not compatible with the development of these eggs.

E.4 Allievi et al. (1994)

Allievi et al. (1994) added KOH and NH_4OH solutions to anaerobically digested, aerobically digested, and a mixture of anaerobic and aerobic biosolids fortified with a high-TKN saline solution. Addition of the solutions increased the pH of the samples to 10.5. Samples were all 10% solids by weight. Next, they stored the samples at 20-25°C (68-77°F) for 60 days. They measured the densities of aerobic bacteria, anaerobic bacteria, fungi, total coliforms, fecal coliforms, and fecal streptococci in each of the six biosolids samples during storage. These alkaline treatments reduced the densities of fecal organisms but had little effect on the nonfecal organisms. The authors considered survival of nonfecal microorganisms to be a positive development, because continued microbial activity in biosolids after destruction of fecal organisms provides protection against subsequent contamination by more lethal microorganisms.

NH_4OH treatment was much more effective than KOH treatment. NH_4OH inactivated fecal coliforms immediately upon dosing and fecal streptococci within 14 days. KOH treatment required 14 days to inactivate the fecal coliforms. It effected only slight reductions in fecal streptococci, about the same amount as fecal coliforms and fecal streptococci densities were reduced in control samples. Further experiments were conducted with the more resistant fecal streptococci.

The authors found that disinfection at pH 4 with phosphoric acid is more rapid than disinfection with ammonia. However, low-pH treatment was not deemed practical because of increased potential for corrosion of spreading equipment and effervescence during acid addition.

Then authors found that ammonia effectively inactivated fecal streptococci only when the pH was above about 10. They also found that ammonia demands increased rapidly for pH values much above 10. The pH 10 condition was therefore judged to be a condition of economic operation, and further experiments were run mainly at pH 10. The pH declined marginally (only about 0.5 units) during storage when the initial pH was near 10.

Further experiments showed that ammonia's disinfection effects were similar at 10°C, 20°C, and 28°C (50°F, 68°F, and 82°F), but that treatment at 5°C (41°F) was ineffective.

The authors also dosed parallel samples of a mixed biosolids with 1) a saline solution of ammonium sulfate and potassium chloride (pH of this experiment not stated); 2) NH_4OH to produce a pH of 10.2; 3) NH_4OH + saline solution at a pH of 9.8; and 4) no chemicals (a control sample). The samples were then stored at 20°C (68°F) for 20 days. Fecal streptococci were inactivated in seven days in the NH_4OH + saline solution biosolids; in 20 days in the NH_4OH biosolids; and not at all in the saline and control biosolids. The authors took more rapid inactivation in the NH_4OH + saline biosolids (at pH of 9.8) compared to the NH_4OH biosolids (at pH of 10.2) as evidence that ammonia has a more powerful effect than pH in inactivating fecal streptococci. This view is supported by the initial experiments that found NH_4OH more effective than KOH.

E.5 Yoshida (1920)

Yoshida (1920) evaluated the inactivation of *A. lumbricoides* by commercial chemicals. He also found that *Ascaris* eggs were unable to develop in urine and they eventually die. He surmised that urine was a more efficient disinfectant at 31°C (88°F) than at 10°C (50°F).

E.6 Fenters et al. (1979)

Fenters et al. (1979) studied the effect of ammonia on Polio I virus (Sabin strain). Ammonia solutions were prepared by adding NH_4OH and/or NH_4Cl to phosphate-buffered saline (PBS) to give ammonia concentrations of 50 mg/L, 100 mg/L, 250 mg/L, 500 mg/L, 750 mg/L, 1000 mg/L, and 1500 mg/L. The solution pH was controlled at 7.0. Penicillin and streptomycin were added to the above and to a control to help contain bacterial multiplication so the tissue culture systems would not become contaminated. Polio I virus was then added to the solutions, and the solutions were divided into 2-mL aliquots in screw-top tubes. The aliquots were then incubated at 35°C (95°F) for 14 days. The aliquots were analyzed for virus after 1 hour, each day for the next nine days and on Days 11 and 14. Test results, expressed as percent inactivation per day, showed inactivation varying linearly from about 45% per day for an ammonium concentration of 0 mg/L to about 60% per day for an ammonia concentration of 1500 mg/L.

The study did not distinguish between the kill caused by the NH_4^+ ion and the kill caused by aqueous NH_3. Total ammonia partitions itself into NH_4^+ and NH_3 according to the relationship $[NH_4^+]/[NH_3] = K\,[H^+]$, where [X] is the activity of component X and $K = 10^{-9.2}$ at 25°C (77 °F).

At pH values in the physiological range (6 to about 8.5) NH_4^+ predominates. However, NH_3, which is the more potent virucide, becomes a significant portion of the mix as the pH increases.

E.7 Cramer et al. (1983)

Cramer et al. (1983) studied the kinetics of poliovirus type 1 (strain CHAT) and bacteriophage f2 in such a way that the effects of OH^- and NH_4^+ could be separated from those of NH_3. Purified virus stocks were placed into solutions of NH_4Cl and control solutions containing NaCl and incubated at 20°C (68°F). Stock ammonia solutions were made by dissolving NH_4Cl in sterile 0.1 M Tris made with ammonia-free water. Working solutions were made by adding appropriate amounts of stock solution to sterile 0.1 M Tris and adjusting the pH to the desired value with 0.5 M NaOH. The purified poliovirus or f2 solutions were added to the ammonia solutions to give initial concentrations of 2.1×10^6 to 2.0×10^7 PFU/mL. The control solutions were prepared in the same way, except that NH_4Cl was replaced by an equivalent amount of NaCl so both solutions would have the same ionic strength.

The authors conducted several experiments including ones where 1) the NH_3 concentration was varied by varying the pH while holding the total ammonia concentration constant; 2) the NH_3 concentration was varied by varying the total ammonia concentration while holding the pH constant; and 3) the NH_3 concentration was held constant and the total ammonia concentration and pH varied.

The tests showed a dose-response relationship between NH_3 and the test organisms. That is, NH_3 was the component primarily responsible for organism death. OH^- had no virucidal effect, other than to raise the pH, which in turn increased the NH_3 concentration. NH_4^+ and Na^+ had similar effects, insignificant compared to NH_3.

Poliovirus was much more susceptible to NH_3 than bacteriophage f2. The log of percent survival versus time curves were linear for both poliovirus and f2 tests, suggesting a rate equation of the form $v = k_0[V][NH_3]^b$, where v is the inactivation velocity, [V] and $[NH_3]$ are the virus and un-ionized ammonia concentrations, respectively, and k_0 and b are temperature- and organism-specific constants.

APPENDIX F

Overview of Subpart D of the Part 503 Regulations

The information in this appendix, which is intended to provide a broad overview of Subpart D of the Part 503 regulation, was taken largely from "Control of Pathogens and Vector Attraction in Sewage Sludge", EPA/625/R-92/013, revised October 1999 (U.S. EPA, 1999). The information in this appendix is not complete. For further details, see the reference.

Biosolids is the term that refers to sewage sludge that has undergone treatment and meets federal and state standards for beneficial use. Beneficial use means application of the biosolids on land, including agricultural land, strip-mined land, forests, plant nurseries, cemeteries, parks, lawns, and gardens. Beneficial use is distinguished from disposal, where biosolids are not reused (e.g., they are placed in dedicated landfills).

F.1 Class A Biosolids

This section describes Class A criteria, the sequential relationship between pathogen reduction and vector attraction reduction, and various alternatives for producing Class A biosolids.

F.1.1 Class A Criteria

Class A biosolids are essentially pathogen free. They can be sold or given away for application on land or in a surface disposal site without any pathogen-related restrictions. There are six alternative methods within the Part 503 regulation for demonstrating Class A pathogen reduction. The objective of these methods is to achieve the following conditions in the end product:

- Either <1000 MPN fecal coliform/g of total biosolids, dry weight basis *or* <3 MPN of *Salmonella sp.*/4 g total biosolids, dry weight basis;
- Enteric viruses: <1 PFU/4 g of total solids, dry weight basis;
- Viable helminth ova: <1 viable helminth ovum/4 g of total solids, dry weight basis.

One of the VAR requirements must also be met when biosolids are applied to land or placed on a surface disposal site. Vectors are living organisms capable of transmitting a pathogen from one organism to another either mechanically (by simply transporting the pathogen) or biologically (by playing a specific role in the life cycle of the pathogen). Likely vectors for sewage sludge pathogens include insects, rodents, and birds. VAR requirements are discussed later in this appendix.

F.1.2 Pathogen Reduction Before or Concurrent with VAR

With two exceptions, Class A pathogen reduction must be accomplished before or at the same time as VAR reduction. Contamination of biosolids with a bacterial pathogen after one of the Class A pathogen reduction alternatives has been conducted may allow bacteria to regrow unless 1) an inhibitory chemical is present, 2) the biosolids are too dry to allow regrowth, 3) little food remains for microorganisms to consume, or 4) an abundant population of non-pathogenic bacteria are present. Nonpathogenic bacteria can repress the growth of pathogens by competing for the food sources and can destroy pathogens through enzymatic reactions.

Biological processes like anaerobic and aerobic digestion can satisfy VAR requirements by minimizing the material that attracts vectors. They also reduce the food supply for pathogens and promote competition for that supply by generating large numbers of non-pathogenic bacteria. If the pathogen-reduction process follows the VAR process, it may eliminate large numbers of non-pathogenic bacteria, leaving pathogens a clear field to regrow. Requiring pathogen reduction to occur before or concurrent with VAR ensures a substantial presence of non-pathogens in the final product.

The exceptions to the requirement of "Class A processes before or concurrent with VAR processes" are heat drying and alkaline treatment. With these processes, the biosolids are either too dry or too chemically inhibited (high pH) to support pathogen regrowth. However, the conditions that prevent regrowth in these exceptions must remain in force until the time the biosolids are disposed, sold, or given away.

F.1.3 Class A Alternatives

Class A biosolids must meet either a fecal coliform limit of <1000 MPN/g of dry solids or a *Salmonella sp.* limit of <3 MPN/4 g of dry solids *and* one of the following six alternatives:

1. *Time and Temperature Requirements.* Contact time and temperatures during that time are specified. They depend primarily upon the material's solids content. Viruses and viable helminth ova are assumed to satisfy Class A criteria in biosolids treated in this manner, and their monitoring is not required.
2. *Alkaline Treatment Requirements.* Biosolids pH must be >12 for at least 72 hours and the biosolids temperature must exceed 53°C (127°F) during that period. Then, the biosolids must be dried to a concentration exceeding 50% solids. Viruses and viable helminth ova are assumed to satisfy Class A criteria in sludge treated in this manner, and their monitoring is not required.
3. *Treatment by Other Processes.* This alternative applies to processes that do not meet the conditions required by Alternatives 1 and 2. Sometimes viruses and helminth ova are not detected in the *untreated* material (feed). In this case, the absence of these pathogens in the treated biosolids does not demonstrate the process can reduce them to the Class A criteria. Therefore, this alternative requires untreated sludge to be monitored for viruses and pathogens. If these pathogens are not detected in the feed sludge, the treated biosolids are assumed to be Class A biosolids until the next monitoring period. Monitoring continues until viruses and/or helminth ova in the untreated material exceed Class A criteria. Then, the treated biosolids are analyzed to see if these pathogens meet Class A criteria. If they do, the treated biosolids meet Class A criteria as long as the treatment process is operated under the

treatment conditions that successfully reduced the pathogens. Then, virus and helminth monitoring is not required.

4. *Unknown Processes.* This alternative applies to processes (such as lagoon storage, air drying, or cake storage) where kill mechanisms are not well understood or there is a lack of control over kill mechanisms. Virus and helminth monitoring are required for each batch of product.

5. *Processes to Further Reduce Pathogens (PFRP).* This alternative provides continuity with 40 CFR Part 257 regulations, which listed the following processes as PRFPs:

   - *Composting.* Time and temperature requirements depend on the type of compost system being operated (within-vessel, static aerated pile, or windrow). Minimum frequency of turning is specified for the windrow method.
   - *Heat Drying.* Residual moisture and biosolids temperature or wet bulb temperature of the exit gas are specified.
   - *Heat Treatment.* Liquid biosolids temperature and contact time are specified.
   - *Beta-ray Irradiation.* Radiation doses are specified.
   - *Gamma-ray Irradiation.* Radiation doses are specified.
   - *Pasteurization.* Contact time and temperature are specified.

Viruses and viable helminth ova are assumed to satisfy Class A criteria in the biosolids treated as specified and their monitoring is not required.

6. *Process Equivalent to a PFRP.* Alternative 6 allows new processes to be determined equivalent to a PFRP. U.S. EPA's PEC provides guidance and recommendation on equivalency to the permitting authority. To be equivalent, a process must consistently reduce pathogens to levels achieved by the listed PFRPs. Equivalency can be site-specific. However, if it can be shown that the process can consistently produce the required pathogen reductions under the variety of conditions that may be encountered across the country, it may qualify for national equivalency. Processes equivalent to PFRPs need not be monitored for viruses and helminth ova unless requirements are stated by the U.S. EPA. To date, the PEC has recommended several processes for equivalency. They include:

   - A two-stage stabilization process from CBI Walker, Inc., Aurora, IL (national equivalency)
   - An autothermal thermophilic digestion process from Fuch Gas and Wassertechnik, Gmbh, Mayen, Germany (national equivalency)
   - A composting process from International Process Systems Inc., Glastonbury, CT (national equivalency)
   - A two-phase thermo-meso feed sequencing anaerobic digestion system from Lyonnaise des Eaux, Le Pecz-Sur-Seine, France (site-specific equivalency)
   - An advanced alkaline-stabilization system with subsequent accelerated drying from N-Viro Energy Systems, Ltd., Toledo, OH (national equivalency)
   - An oxyozonation system from Synox Corporation, Jacksonville, FL (national equivalency)

- ❑ Storage lagoon/air drying system from MWRD of Greater Chicago (conditional site-specific equivalency)

F.2 Class B Biosolids

Unlike Class A biosolids, which are essentially pathogen free, Class B biosolids may contain some pathogens as defined by the Part 503 regulation. Class B pathogen requirements can be met in three ways. The implicit objective of all three alternatives is to ensure that pathogenic bacteria and enteric viruses are reduced in density, as demonstrated by a fecal coliform density of <2 million MPN/g dry solids in the treated product. Viable helminth ova are not necessarily reduced in Class B biosolids.

Site restrictions that restrict crop harvesting, animal grazing, and public access must be applied for a certain period of time to allow environmental factors to further reduce pathogens, particularly helminth ova.

Class B requirements apply to bulk biosolids that are land applied to agricultural areas, forests, reclamation sites, and various controlled-access sites. Biosolids that are placed on surface disposal sites must also meet Class B criteria, unless the active units on which the biosolids are placed are covered over after each working day. Class B biosolids cannot be given away or sold in bags or other containers.

Class B biosolids must also meet one of the VAR requirements. The choice of VAR reduction alternatives may affect the duration of site restrictions.

F.2.1 Class B Alternatives

Class B alternatives are as follows:

1. *Monitoring Fecal Coliforms.* Alternative 1 requires seven samples be taken and that the geometric mean of these samples be <2 million MPN/g of dry solids. Each of the samples should be taken at the same point in the process so that treatment of each sample is equal.
2. *Use of a Process to Significantly Reduce Pathogens (PSRP).* These processes have demonstrated the capability to achieve at least a 2-log reduction in fecal coliform. The processes are:
    - ❑ *Aerobic Digestion.* Mean cell residence times shall be between 40 days at a temperature of 20°C (68°F) and 60 days at a temperature of 15°C (59°F).
    - ❑ *Air Drying.* Biosolids must dry for a minimum of three months. During two of the three months the average daily temperature must be above 0°C (32°F).
    - ❑ *Anaerobic Digestion.* Values for mean cell residence time shall be between 15 days at temperature of 35°C (95°F) and 60 days at a temperature of 20°C (68°F).
    - ❑ *Composting.* Using within-vessel, static aerated pile, or windrow composting, the temperature of the biosolids should exceed 40°C (104°F) or more for at least five days. For a four-hour period during the five-day period, the compost temperature should exceed 55°C (131°F).
    - ❑ *Lime stabilization.* Enough lime must be used to raise the pH to 12 for at least two hours.

3. *Processes Equivalent to a PSRP.* Sewage sludge treated by any process determined to be equivalent to a PSRP are considered to be Class B biosolids. The PEC has recommended two processes as equivalent to PSRP:
   - ❑ An oxyozonation system from Synox Corporation, Jacksonville, FL.
   - ❑ An alkaline stabilization process from N-Viro Energy Systems, Ltd., Toledo, OH.

F.3 VAR Requirements

VAR is accomplished by one of the following three methods:

- ♦ Biological processing to break down VSS to reduce available food for microbial activity and odor production potential.
- ♦ Chemical or physical conditions that stop microbial activity.
- ♦ Physical barriers between vectors and the biosolids.

Stable biosolids are generally defined as biosolids which have little food available for microbial activity and are not highly odorous. Stable biosolids will normally meet VAR requirements. However, biosolids that meet VAR requirements may not necessarily be stable.

VAR requirements must be satisfied separately from pathogen-reduction requirements. Class A requirements mandate that pathogen reduction must be conducted before or concurrent with VAR operations to prevent regrowth of pathogenic bacteria. This condition is not required for Class B biosolids (i.e., VAR Options 9 or 10, see below).

There are 12 options available to satisfy VAR requirements. The first eight options are designed to reduce sewage sludge's attractiveness to vectors. The last four options are designed to prevent the vectors from contacting sewage sludge.

1. *VSS Reduction.* This option pertains to biological treatment systems, such as digestion. VSS concentrations must be decreased by at least 38% during sludge treatment, which can include digestion plus any add-on processes, such as lagoon storage, air drying, or cake storage.
2. *Additional Digestion of Anaerobically Digested Biosolids.* This option pertains to biosolids that are already considerably reduced in VSS content via anaerobic digestion but cannot meet Option 1 requirements. Option 2 requires that <17% additional VSS loss to occur during bench-scale batch anaerobic digestion of the biosolids at temperatures of 30-37°C (86-99°F) for 40 days.
3. *Additional Digestion of Aerobically Digested Biosolids.* This option pertains to biosolids that are already considerably reduced in VSS content before aerobic digestion, such as biosolids from an extended aeration system. Such biosolids may have difficulty satisfying Option 1 requirements (38% VSS reduction in the biosolids treatment system). Option 3 requires that <15% additional VSS loss occur during bench-scale batch aerobic digestion of the biosolids at 20°C (68°F) for 30 days. This test applies to biosolids with concentrations below 2% solids. Biosolids of higher concentration must be diluted to <2% solids before running this test.

4. *Specific Oxygen Uptake Rate (SOUR) for Aerobically Digested Biosolids.* This option applies to aerobically digested biosolids of concentrations <2% solids that have been aerobically digested or partially digested at temperatures between 10 °C and 30°C (50 °F and 86°F). Option 4 requirements are satisfied when the SOUR is <1.5 mg of oxygen/hour/g of total sewage solids (dry weight basis) at 20°C (68°F).

5. *Aerobic Processes at Greater than 40°C (104°F).* This option applies primarily, but not exclusively, to composting. The biosolids must be aerobically treated for at least 14 days. During this time, the solids temperature must be over 40°C (104°F) and the average temperature higher than 45°C.

6. *Addition of Alkali.* Option 6 requirements are satisfied when sufficient alkali (e.g., lime, fly ash, kiln dust, and wood ash) is added to raise the pH to 12 (as measured at 25°C) for 2 hours and to maintain it at 11.5 for 22 hours thereafter. A pH temperature correction is presented for sample temperatures other than 25°C.

7. *Moisture Reduction of Biosolids Containing No Unstabilized Solids.* This option pertains to biosolids treated by aerobic or anaerobic processes (i.e., biosolids that contain no primary or raw solids). The biosolids must have a solids concentration that exceeds 75% before they are mixed with any other materials. The 75% or more solids concentration must be achieved by removing water, not by adding inert materials.

8. *Moisture Reduction of Sewage Sludge Containing Unstabilized Solids.* This option applies to sludge that contains unstabilized solids from primary treatment. This sludge must have a solids concentration exceeding 90% before it is mixed with any other materials.

9. *Injection.* This option requires injection below the earth's surface. It puts an earthen barrier between the biosolids and the vectors. No significant amount of biosolids can be present on the surface within one hour after injection. Class A biosolids (which in some cases may contain relatively few nonpathogenic bacteria to suppress regrowth of pathogens) must be injected within eight hours after discharge from the pathogen-reduction process. The eight-hour restriction does not apply to Class B biosolids, which tend to have significant populations of non-pathogenic bacteria.

10. *Incorporation of Biosolids into the Soil.* This option applies to biosolids applied to the land (e.g., for agricultural purposes) or placed on a surface disposal site. The biosolids must be incorporated into the land (by plowing or other forms of mixing the biosolids and soil) within six hours after application. As with Option 9, Class A biosolids must be incorporated into the soil within eight hours from discharge from the pathogen-reduction process.

11. *Covering Biosolids.* Under this option, biosolids placed on a surface disposal site must be covered with soil or other material at the end of each operating day.

12. *Raising the pH of Domestic Septage.* This option applies only to domestic septage applied to agricultural land, forests, reclamation sites, or surface disposal sites. The pH must be raised to at least 12 by alkali addition and held there for 30 minutes without adding more alkali. Every truckload of septage must be monitored to ensure it meets the pH standard.

APPENDIX G

CONDITIONS TO PRODUCE CLASS A BIOSOLIDS

Appendix G develops operating guidelines for lagoon storage, air drying, cake storage, and combined lagoon storage/air drying systems for producing Class A biosolids. The guidelines describe conditions for reducing the densities of helminth ova to the Class A criterion (1 viable ovum/4 g dry solids). Helminth ova are reportedly the hardiest of the regulated pathogens found in municipal sludge. Reduction of viable helminth ova to levels stipulated in the Part 503 regulations likely ensures destruction of the other regulated pathogens. The destruction of helminth ova thus dictates processing times for lagoon storage, air drying, and cake storage processes.

The guidelines assume all low-technology processes are preceded by effective biological digestion. The literature suggests that anaerobic digestion is more effective than aerobic digestion in this pretreatment role; that is, low-technology processes inactivate pathogens better when combined with anaerobic digestion than they do combined with aerobic digestion. The higher concentrations of ammonia and organic acids in anaerobically digested biosolids, which carry through the low-technology processes, may account for part of this result.

G.1 Lagoon Storage

Kaneshiro and Stern (1985) showed that viable *Ascaris* ova can be reduced to Class A levels by storage for extended periods at temperatures over 20°C (68°F). They also showed that inactivation at 4°C (39°F) is quite slow. Table G-1 shows selected data from their report.

The investigators tested the decline of *Ascaris* ova that had been spiked into anaerobically digested or aerobically digested biosolids in a two-phase program:

- *Phase I* Biosolids samples were spiked after they had been digested.
- *Phase II* Sludge samples were spiked before they had been digested.

Storage at 25°C (77°F) reduced densities of viable helminth ova to the Class A criterion, with $\log_{10}$ reductions in the range 3.7 to 4.4. Storage at 4°C (39°F) for 25 months did not reduce viable helminth ova to the Class A criterion.[1] Log_{10} density reductions ranged from 0.8 to 1.6 at 4°C (39°F). The data show that some ova were completely destroyed and disappeared and some were simply inactivated.

[1] Biosolids concentrations in the Kaneshiro and Stern experiments were about 3%. Therefore, multiply Table G-1 densities by 1.33 to obtain approximate number of ova/4 g dry solids. This calculation will allow comparison of helminth densities to the Class A criterion (1 viable ovum/ 4 g dry solids).

Table G-1. Decline of *Ascaris* Ova with Time in Stored Biosolids Samples.

Prior Digestion	Temperature (°C)	Phase	Exposure Time (Months)	Helminth Ova (no./100 g Wet Solids) Total	Viable
Anaerobic	25	I	0	7800	7400
			16	3367	0
		II	0	5300	5000
			16	45	0
Aerobic	25	I	0	8300	7700
			19	102	0.3
		II	0	4800	4500
			16	52	0
Anaerobic	4	I	0	7800	7440
			25	1574	1,247
		II	0	5300	5000
			25	177	135
Aerobic	4	I	0	8300	7700
			25	1772	672
		II	0	4800	4500
			25	236	195

Source: Kaneshiro and Stern, 1985

Reimers et al. (1989) observed the decline of spiked helminth ova in several pilot lagoons holding anaerobically digested biosolids.

- *Pilot lagoons*: 1.8 m x 1.8 m x 3.0 m (5.9 ft x 5.9 ft x 5.9 ft) deep, volume about 7.5 m^3 (1981 gal)
- *Biosolids/tube lagoons*: 2.4 m (7.8 in.) long by 10.2 cm (4 in.) diameter, spiked with much higher concentrations of *Ascaris* ova than the pilot lagoons. The tubes were lowered into the pilot lagoons and therefore experienced the same temperature regimes.
- *Biosolids columns*: 3.66 m (12 ft) long by 0.3 m (1 ft) diameter. These columns were loaded with the same biosolids as the pilot and biosolids/tube lagoons.

Tests were conducted with anaerobically digested biosolids from New Orleans, Louisiana, and Port Aransas, Texas. The pilot lagoon and biosolids/tube/lagoon tests were conducted in those locations. The biosolids column tests were conducted in New Orleans. Temperatures of the lagoons varied. Table G-2 presents selected results from these studies.

Table G-2. Decline of *Ascaris* Ova with Time in Pilot Lagoons.

Biosolids Source	Description	Temperature (°C)	Exposure Time	Helminth Ova (no./10 g Dry Solids) Total	Viable
New Orleans	Pilot lagoon	Avg. 20.7, range 13-29	24 hours	2,335	194
			12.4 months	718	0
	Biosolids/tubelagoon	Avg. 20.7, range 13-29	3 months	120,000	19,900
			12.2 months	63,100	0
	Biosolids column	Avg. 25, range 20-27	3.3 months	11,493	10,625
			15.2 months	9,006	0
Port Aransas	Pilot lagoon	Avg. 24.7 range 19-29	24 hours	1,019	64
			6.5 months	896	0.2
	Biosolids/tube lagoon	Avg. 24.7, range 19-9	3.4 months	28	1
			5.4 months	99	0
	Biosolids column	Avg. 25, range 20-27	3.3 months	1,571	1,064
			12 months	1,141	0.7

Source: Reimers et al., 1989

Table G-2 shows that viable helminth ova densities were reduced to the Class A criterion in 12 to 15 months in New Orleans biosolids and in 5.4 to 12 months in Port Aransas biosolids. Log_{10} reductions in viable helminth ova ranged from 2.3 to 4.3 in New Orleans biosolids and from 1.0 to 3.2 in Port Aransas biosolids. Some ova were completely destroyed but that more ova were inactivated. The investigators speculated that the presence of petrochemical wastes in the Port Aransas biosolids was responsible for the more rapid decline in helminth ova. They believed the low heminth densities in the Port Aransas biosolids/tube lagoon were the result of a failure to spike the biosolids.

To summarize, the ability of a lagoon system to produce Class A product depends primarily on the density of viable helminth ova in lagoon feed, biosolids temperature, and biosolids hold time (time without any addition of fresh feed). Other factors (method of prior biosolids digestion and chemical factors) may also influence performance. A lagoon operating at 25°C (77° F) or above with preceding anaerobic digestion would appear capable of producing Class A biosolids within 16 months of pure storage time under most conditions conceivable. If the log_{10} reduction performance obtained in the Kaneshiro and Stern experiments could be maintained at plant scale (there is no certainty that it could be), lagoons operating at 4°C might produce Class A biosolids within 25 months if the density of viable helminth ova in lagoon feed were reasonably low (5 to 10 ova/4 g dry solids).

There is less information about the rates of inactivation of the *Ascaris* ova (the most-resistant pathogen) in the temperature range between 4 °C and 25°C (39 °F and 77°F). Therefore, definitive storage-time guidance cannot be given for this temperature range. Such information is needed, since most storage lagoons operate in this range.

G.2 Air Drying and Cake Storage

This project's literature review suggests that air drying can reduce the densities of viable helminth ova to the Class A criterion by desiccation and by exposure to temperature over time.

G.2.1 Desiccation-dominated Drying

Cram (1943) and Baxter et al. (1983) found it necessary to reduce biosolids moisture contents to below 5% to achieve complete inactivation of helminth ova. However, operators of most U.S. air drying plants will be unable to dry their biosolids to 5% moisture levels or less because of rainfall or humidity constraints. Operators in hot, arid regions like the southwestern United States may be the exception. Even those who can achieve such low moisture levels may be unwilling to do so because of dusting problems.

Drying to very low moisture levels can be accomplished fairly quickly in arid regions, often within two to four months. However, reduction to low moisture levels does not guarantee Class A results, especially if biosolids temperatures have not been raised substantially. Wilson and Co. (1989) observed that relatively rapid drying (6 weeks) in low windrows at Roswell, New Mexico, left high densities of salmonellae and fecal coliform even when the product moisture level dropped to 10%. Windrow temperatures were consistently cooler than the soil temperature during the drying period discussed. The reduction in fecal coliform and salmonellae densities correlated better with time on the bed than any other parameter, including final product moisture content. Drying on sand beds produced greater reductions in bacterial densities under the same environmental conditions. In arid climates it is possible to reach very high temperatures in the thin layer remaining on a sand bed at the end of the drying cycle.

The Roswell experience shows drying time and drying temperature still play significant roles in desiccation-oriented drying systems. However, no time and temperature guidance can be given for these systems at this time.

G.2.2 Time and Temperature-based Drying

When drying to low moisture levels is not an option, operators must hold the biosolids long enough at high enough temperatures to ensure inactivation of helminth ova. Effective treatment was found under the following conditions:

- *Warm summers, mild winters* (Austin, Texas). Air drying anaerobically digested biosolids on an impervious pad reduced densities of viable helminth ova from >12 ova/4 g dry solids to <1 ovum/4 g dry solids in about 250 days. During this time, the biosolids were turned three times per week and dried from an initial moisture content of about 95% to a final moisture content of about 75% (Lucero-Ramirez and Malina, 1998).
- *Warm summers, cold winters* (five towns in Utah). Cake storage of aerobically and anaerobically digested biosolids on impervious pads reduced densities of spiked ova of *A. suum* from initial values ranging from 640 ova/4 g dry solids to 16,000 ova/4 g dry solids to the detection limit of the analysis used (5 ova/4 g dry solids) in about 330 days when the biosolids were turned at least once a month. The biosolids dried from initial moisture concentrations of 40-70% to final moisture contents of <25% (Ahmed and Sorensen, 1997). The investigators appear to have reported total intact helminth ova densities, not viable helminth ova densities. Since the latter should be significantly less than the former, it is likely the final product satisfied Class A limits for viable helminth ova.

This information suggests the following conservative design parameters for air drying and cake storage systems on impervious pads in temperate climates:

- Air drying or cake storage for at least 250 days for climates with warm summers and mild winters and for at least 350 days for climates with warm summers and cold winters.
- Drying periods must encompass at least one summer.

- Pile or windrow moisture contents should remain above 40% for most of the storage or drying periods to promote continued biological activity.
- Turning as required to control the windrow moisture content and temperature.

G.3 Combined Lagoon Storage and Air Drying

Lagoons and air drying are often used in sequence to produce Class A biosolids. Because organisms exit the lagoons in a debilitated state they are vulnerable to the altogether different environment they encounter in the air drying processes. Therefore, this combination is probably more powerful and likely to take less time than lagoon storage, air drying, or cake storage alone, even when temperature conditions are the same for all.

Table G-3 presents information from treatment plants operating combined lagoon storage/air drying systems. All are producing Class A biosolids. All employ mesophilic anaerobic digestion as the first sludge treatment step. Chicago has monitored helminth ova reduction throughout its treatment train to develop a data base to support its application for PFRP equivalency. The other systems monitor pathogens only in the dried product. Chicago has very low levels of viable helminth ova. The levels of viable helminth ova are below the Class A limit even its raw (untreated) sludge.

The solids detention time in Chicago's digesters is 20 days. San Jose and Goleta have much longer detention times. The average temperature of San Jose's lagoons ranges from 15-18°C (59-64°F). Mean temperatures in the sludge layers of Chicago's lagoons are in the 13-15°C (55-59°F) range, with low temperatures of 6°C (43°F) and high temperatures of 24°C (75°F). Goleta does not report lagoon temperatures. Chicago's drying beds are turned an average of three times a week, Goleta's more frequently. San Jose turns its drying beds several times in an approximately 12- to 20-week drying cycle.

The Goleta Sanitary District does not operate its lagoons to provide a definitive unfed storage time. However, when Goleta dredges biosolids from a lagoon, that lagoon is typically not being fed digested biosolids, and there is often several weeks or more between the last feeding of a lagoon and the dredging of biosolids from that lagoon to the air drying pad area.

Table G-3. Treatment Plants with Combined Lagoon Storage/Air Drying Systems.

Location	Anaerobic Digestion SRT (Days)	Lagoon Storage: Unfed Storage Time (Months)	Lagoon Storage: Log Reduction, *Ascaris* Ova	Air Drying: Drying Time (Weeks)	Air Drying: Log Reduction, *Ascaris* Ova	Reference
Chicago, IL						
Stickney, LSSPT	20	18 minimum, 24 actual	0.95	9 (average)	0.52	Tata et al., 2000
Stickney, HSSPT	20	18 minimum, 19 actual	0.65	10 (average)	0.62	
Calumet, LSSPT	20	18 minimum, 30 actual	1.67	9 (average)	1.2	
Calumet, HSSPT	20	18 minimum, 19 actual	2.57	10 (average)	0.69	
San Jose, CA	Up to 40 days	24	NM[a]	Approx. 12 to 20	NM[a]	Garner, 2004
Goleta Sanitary District, CA	50	Limited (see text)	NM	6 to 12	NM	Salt, 2004

[a] NM = not measured

While San Jose and Goleta produce Class A biosolids through final product pathogen testing, the lack of information about viable helminth ova in lagoon feed prevents conclusions about their systems' capability to eliminate viable helminth ova. The MWRD of Greater Chicago has used a statistical approach to predict whether or not its sludge treatment systems could satisfy Class A criteria in the event of periodic surges in helminth and *Ascaris* ova of up to 10,000 times their average loadings (Tata et al., 2000a). It concluded that its air-dried product would continue to achieve Class A standards when these events occurred.

Summarizing, Chicago and San Jose employ unfed lagoon storage times of 18 and 24 months, respectively. The Goleta Sanitary District provides limited and variable time for unfed lagoon storage. The SRT for the Goleta digesters is unusually long, however, suggesting the digesters are performing extra pathogen kill that would normally be achieved in downstream lagoons. The combined systems' air drying times for all three agencies are also not as long as recommended in this report for systems operating singly (i.e., not in combination). Drying systems in all three locations appear to be run only long enough to produce a dry product.

The Chicago model appears to be the best available for combined lagoon/air drying systems.

Glossary of Terms

Indicator organism	An organism that is numerous and easily tested (such as fecal coliform) whose presence indicates the presence of less numerous and more difficult-to-test pathogenic organisms
Beneficial use	Application of biosolids on land, including strip-mined land, forests, plant nurseries, cemeteries, parks, lawns, and gardens
Biosolids	Sewage sludge that has undergone treatment and meets federal and state standards for beneficial use
Most probable number	A statistics-based estimate of the number of organisms present. It is based on the number of positive and negative results obtained when testing multiple portions in a geometric series for the organism of interest.

References

Ahmed, A.U.; Sorensen, D.L. Kinetics of Pathogen Destruction During Storage of Dewatered Biosolids. *Water Envir. Res.* **1995,** 67, 143-150.

Ahmed, A.U.; Sorensen, D.L. Autoheating and Pathogen Destruction During Storage of Dewatered Biosolids with Minimal Mixing. *Water Envir. Res.* **1997**, 69, 81-94.

Al-Ghazali, M.R.; Al-Azawi, S.K. Storage Effects of Sewage Sludge Cake on the Survival of L. *monocytogenes. Journal of Applied Bacteriology* **1988,** 65, 209-213.

Allievi, L.; Colombi, A.; Calcaterra, E.; Ferrari, A. Inactivation of Fecal Bacteria in Sewage Sludge by Alkaline Treatment.*Bioresource Technology* **1994**, 49, 25-30.

Amin, O.M. Pathogenic Micro-organisms and Helminths in Sewage Products, Arabian Gulf, Country of Bahrain. *American Journal of Public Health* **1988**, 78 (3), 314-315.

Baxter, J.C.; Johnson, D.; Burge, W.D. *Effects of Grazing Cattle on Forage Fertilized With Sewage Sludge;* Rept. No. EPA1600/2-83-050(NTIS #PB 83-170589); Municipal Environmental Research Laboratory, U.S. EPA, 1983.

Bergman, W.A.; Rohloff, G.M. Complete Costs for Biosolids Processing at the Metropolitan Water Reclamation District of Greater Chicago; Alexandria, VA, 1996; Water Environment Federation.

Bitton, G.; Farrah, S.R.; Overman, A.R.; Gifford , G.E.; Pancorbo, O.C.; Charles, J.M.. *Fates of Viruses Following Application of Municipal Sludge to Land.* Unpublished manuscript. See Pedersen (1981) for a summary of this work.

Bowman, D.D.; Little, D.D.; Reimers, R.S. Precision and Accuracy of an Assay to Detect *Ascaris* Eggs in Various Biosolids Matrices. *Water Research* **2003**, 37,2063-2072.

Cram, E.B. The Effect of Various Treatment Processes on the Survival of Helminth Ova and Protozoan Cysts in Sewage. *Sewage and Industrial Wastes* **1943**, 15(11), 1119-1138.

Cramer, W.N.; Burge, W.D.; Kawata, K. Kinetics of Virus Inactivation by Ammonia. *Applied and Environmental Microbiology* **1983**, 45 (3),760-765.

Farrah, S.R.; Bitton, G.; Hoffmann, E.M.; Lanni, O.; Lutrick, M.C.; Bertrand, J.E. Survival of Enteroviruses and Coliform Bacteria in a Sludge Lagoon. *Applied and Environmental Microbiology.* **1981**, 41(2), 459-465.

Fenters, J.; Reed, J.; Lue-Hing, C.; Bertucci, J. Inactivation of Viruses by Digested Sludge Components. *Journal of the Water Pollution Control Federation* **1979**, 51 (4), 689-694.

Garner, R. City of San Jose, CA. Personal Communication, May 21, 2004.

Gibbs, R.A.; Hu, J.C.; Ho, G.E.; Phillips, P.A; Unkovich, I. Pathogen Die-off in Stored Wastewater Sludge. *Water Science and Technology*, **1995**, 31(5,6), 91-95.

Ghiglietti, R.; Rossi, P.; Ramsan, M.; Colombi, A. Viability of *Ascaris Suum, Ascaris Lumbricoides,* and *Trichuris Muris* Eggs to Alkaline pH and Different Temperatures. *Parasitologia* **1995**, 37, 229-232.

Jepsen, S.E.; Krause, M.; Gruttner, H. Reduction of Fecal Streptococcus and *Salmonellae* by Selected Treatment Methods for Sludge and Organic Waste. *Prog. Water Technol.* **1997**, 36, 203.

Kaneshiro, E.S.; Stern, G. 1985. *Survival of Parasite Eggs in Stored Sludge.* U.S. EPA Report. No. EPA 600/1-87-014 (NTIS No. PB 88-154273/AS). U. S. Environmental Protection Agency, Cincinnati, OH, 1985.

Lucero-Ramirez, B.; Malina, J.F. Jr. The Effect of Standard Municipal Sludge Treatment Processes on the Reduction of Indicator and Pathogenic Organisms. *Proceedings of WEFTEC 98, Water Environment Federation 71st Annual Conference*, Orlando, FL, 1998.

Lucero-Ramirez, B. The Effects of Time and Temperature on the Fates of Pathogens and Indicator Bacteria During Municipal Wastewater Sludge-Mesophilic Anaerobic Digestion, Air Drying, and Composting. Ph.D. Thesis, Department of Civil Engineering, University of Texas, Austin, TX, 2000.

Lue-Hing, C; Sedita, S.J.; Tata, P.; Bertucci, J.J.; Kambhampati, C.R.; Zenz, D.R.; Knafl, G.J. *Final Report on Certification of the Sludge Processing Trains (SPT) of the Metropolitan Water Reclamation District of Greater Chicago (District) as Equivalent to process to Further Reduce Pathogens (PFRP);* Metropolitan Water Reclamation District of Greater Chicago; Chicago, IL; 1999.

Mbela, K.K. Evaluation of Temperature Effects on Inactivation of *Ascaris* Eggs in Both Aerobic and Anaerobic Digestion Process. M. S. Thesis, Tulane University, New Orleans, LA, 1989.

McHaney, S.X; Rosenblum, E.; Babatola, A.; Ihrke, D. Cost-Effective Biosolids Dewatering and Pathogen Reduction by Solar Drying at the San Jose/Santa Clara Water Pollution Control Plant. In *Proceedings of Management of Water and Wastewater Solids for the 21st Century: A Global Perspective*, Washington D.C. June 19-22, 1994.

Miami-Dade Water and Sewer Department. Data from sludge/biosolids air drying tests conducted at South Plant over the period September 1997 to January 1998, 1998.

O'Donnell, C.J.; Meyer, K.B.; Jones, J.V.; Benton, T.; Kaneshiro, E.S.; Nichols, J.S.; Schaefer, F.W. III. Survival of Parasite Eggs Upon Storage in Sludge. *Applied and Environmental Microbiology.* **1984**, 48(3), 618-625.

Pedersen, D.C. Sludge Storage/Lagooning. *Density Levels of Pathogenic Organisms in Municipal Wastewater Sludge: A Literature Review*. Rept. No. EPA 600/2-81-170 (NTIS no.PB82102286). U. S. Environmental Protection Agency: Cincinnati, OH; 1981; 153-169.

Phillipp, W.; Lang, A.; Strauch, D. Bacteriological and Parasitological Investigations on the Influence of Filter Beds Covered with Reed on the Survival of Salmonellae and *Ascaris* Eggs. *Inactivation of Microorganisms in Sewage Sludge by Stabilization Processes*, Elsevier Appl. Sci. Pub. Ltd. (Elsevier Sci. Pub. Co., Inc., New York), 1985, 168-188

Ponugoti, R.P; Dahab, M.F.; Surampalli, R. *Effects of Different Biosolids Treatment Systems on Pathogen and Pathogen Indicator Reduction.* Water Environment Research **1997,** 69(7), 1195-1206.

Ramey, R.W.; Anderson, T.; Oerke, D. 1996. Using a Low-Cost, Low-Tech Air Drying/Storage Process to Produce Class "A" Biosolids. In *Proceedings of the 10th Annual Residuals and Biosolids Management Conference*, Denver, CO, August 18-21, 1996.

Reimers, R.S.; Little, M.D.; Englande, A.J. Jr; McDonnell, D.B.; Bowan, D.D.; Hughes, J.M. *Investigation of Parasites in Sludges and Disinfection Techniques.* Cooperative Agreement No. CR-806711-01, Toxicology and Microbiology Division, Health Effects Research Laboratory, Cincinnati, OH, 45268, 1985. (Published by NTIS in 1986. NTIS Report No. PB86:135407.)

Reimers, R.S.; Little, M.D.; Akers, T.G.; Henriques, W.D.; Badeaux, R.C.; McDonell, D.B.; Mbela, K.K. Chapter 6, Results. *Persistence of Pathogens in Lagoon Stored Sludge.* Rept. No. EPA/600-2-89/015; U.S. Environmental Protection Agency: Cincinnati, OH, 1989, 33-69.

Reimers, R.S.; Schafer, P.L.; Leftwich, B.; Bowman, D.B.; Tata, P.; Atique, M.M. Factors Affecting Lagoon Storage Disinfection of Biosolids.In *Proceedings of WEF/AWWA/CWEA Joint Residuals and Biosolids Management Conference Biosolids 2001 "Building Public Support"*, 2001.

Salt, J. Goleta Sanitary District, Goleta, CA. Personal communication, May 19, 2004.

Sattar, S.Y.; Westwood, J.C.N. Recovery of Viruses from Field Samples of Raw, Digested, and Lagoon-Dried Sludges. *Bulletin of World Health Organization* **1979**, 57 (1), 105-108

Stern, G.; Farrell, J.B. Sludge Disinfection Techniques. In *Proc. Of Nat. Conf. on Composting Municipal Residues and Sludges*, Washington D.C., Pub. Information Transfer, Inc.: Rockville, MD, 1977; 1117-1122.

Tata, P.; Lue-Hing, C.; Bertucci, J.; Sedita, S.J.; Knafl G.J. Class A Biosolids Production by a Low-Cost Conventional Technology. *Water Envir. Res.* **2000**, 72(4), 413-422.

Tata, P.; Lue-Hing, C.; Knafl, G.J. Statistical Evaluation of Pathogen Inactivation for a Conventional Low-Cost Technology Class A Biosolids Process. *Water Envir. Res.* **2000a.** 72(4), 423-431.

Traub, J.L. Letter from Jo Lynn Traub, Director, Water Division, U.S. Environmental Protection Agency, Chicago, IL to Jack Farnan, General Superintendent, Metropolitan Water Reclamation District of Greater Chicago, Chicago, IL, received June 20, 2002.

U.S. EPA.Storage. *Process Design Manual for Sludge Treatment and Disposal*; EPA 625/1-79-011, U.S. Environmental Protection Agency, Municipal Environmental Research Laboratory: Cincinnati, OH, 1979;15-49.

U.S. EPA. *Policy on Municipal Sludge Management*; 40 FR 24358, June 12, 1984.

U.S. EPA. *Part 503-Standards for the Use or Disposal of Sewage Sludge, Rules and Regulations.* Federal Register, **1993**, (32)58.

U.S. EPA. *Control of Pathogens and Vector Attraction in Sewage Sludge.* EPA/625/R-92/013, revised October 1999; Office of Research and Development, National Risk Management Laboratory, Center for Environmental Research Information, United States Environmental protection Agency: Cincinnati, OH, 1999; 150-172.

Veerannan, K.M. Some Experimental Evidence on the Viability of *Ascaris lumbricoides* Ova.Curr. Science **1977**, 46(11), 386-387.

Ward, R.L.; Ashley, C.S. Identification of the Virucidal Agent in Wastewater Sludge. *Applied and Environmental Microbiology* **1977**, 33 (4), 860-864.

Ward, R.L.; Ashley, C.S. Inactivation of Enteric Viruses in Wastewater Sludge Through Dewatering by Evaporation. *Applied and Environmental Microbiology* **1977**, 34 (5), 564-570.

Wilson & Co., Engineers. *Innovative Sludge Drying Study, 6th Quarterly Report, City of Roswell, New Mexico;* EPA Project No. C-35-1052-10. Wilson & Co., Engineers and Architects, Albuquerque, NM., 1989.

Wilson & Co., Engineers. *Innovative Sludge Drying Field Test Final Report, to City of Roswell, New Mexico;* EPA Project No. C-35-1052-10. Wilson & Co., Engineers and Architects, Albuquerque, NM., 1990.

Yoshida, S. On the Resistance of Ascaris Eggs. *Journal of Parasitology* **1920**, 6(2), 132-139.

WASTEWATER UTILITY

Alabama
Montgomery Water Works & Sanitary Sewer Board

Alaska
Anchorage Water & Wastewater Utility

Arizona
Gila Resources
Glendale, City of, Utilities Department
Mesa, City of
Peoria, City of
Pima County Wastewater Management

Arkansas
Little Rock Wastewater Utility

California
Calaveras County Water District
Central Contra Costa Sanitary District
Contra Costa Water District
Crestline Sanitation District
Delta Diablo Sanitation District
Dublin San Ramon Services District
East Bay Dischargers Authority
East Bay Municipal Utility District
El Dorado Irrigation District
Fairfield-Suisun Sewer District
Fresno Department of Public Utilities
Irvine Ranch Water District
Las Virgenes Municipal Water District
Livermore, City of
Lodi, City of
Los Angeles, City of
Los Angeles County, Sanitation Districts of
Napa Sanitation District
Orange County Sanitation District
Palo Alto, City of
Riverside, City of
Sacramento Regional County Sanitation District
San Diego Metropolitan Wastewater Department, City of
San Francisco, City & County of
San Jose, City of
Santa Barbara, City of
Santa Rosa, City of
South Bayside System Authority
South Coast Water District
South Orange County Wastewater Authority
Sunnyvale, City of
Union Sanitary District
West Valley Sanitation District

Colorado
Boulder, City of
Colorado Springs Utilities
Littleton/Englewood Water Pollution Control Plant
Metro Wastewater Reclamation District, Denver

Connecticut
The Mattabasset District
New Haven, City of, WPCA

District of Columbia
District of Columbia Water & Sewer Authority

Florida
Broward, County of
Fort Lauderdale, City of
Gainesville Regional Utilities
JEA
Miami-Dade Water & Sewer Authority
Orange County Utilities Department
Orlando, City of
Reedy Creek Improvement District
Seminole County Environmental Services
St. Petersburg, City of
Stuart Public Utilities
Tallahassee, City of
Tampa, City of
Toho Water Authority
West Palm Beach, City of

Georgia
Atlanta Department of Watershed Management
Augusta, City of
Clayton County Water Authority
Cobb County Water System
Columbus Water Works
Fulton County
Gwinnett County Department of Public Utilities
Savannah, City of

Hawaii
Honolulu, City and County of

Illinois
American Bottoms Wastewater Treatment Plant
Greater Peoria Sanitary District
Kankakee River Metropolitan Agency
Metropolitan Water Reclamation District of Greater Chicago
North Shore Sanitary District
Wheaton Sanitary District

Iowa
Ames, City of
Cedar Rapids Wastewater Facility
Des Moines Metro Wastewater Reclamation Authority
Iowa City, City of

Kansas
Johnson County Unified Wastewater Districts
Unified Government of Wyandotte County/Kansas City, City of

Kentucky
Louisville & Jefferson County Metropolitan Sewer District

Louisiana
Sewerage & Water Board of New Orleans

Maine
Bangor, City of
Portland Water District

Maryland
Anne Arundel County Bureau of Utility Operations
Howard County Department of Public Works
Washington Suburban Sanitary Commission

Massachusetts
Boston Water & Sewer Commission
Upper Blackstone Water Pollution Abatement District

Michigan
Ann Arbor, City of
Detroit, City of
Holland Board of Public Works
Lansing, City of
Owosso Mid-Shiawassee County WWTP
Saginaw, City of
Wayne County Department of Environment
Wyoming, City of

Minnesota
Rochester, City of
Western Lake Superior Sanitary District

Missouri
Independence, City of
Kansas City Missouri Water Services Department
Little Blue Valley Sewer District

Nebraska
Lincoln Wastewater System

Nevada
Henderson, City of

New Jersey
Bergen County Utilities Authority
Passaic Valley Sewerage Commissioners

New York
New York City Department of Environmental Protection
Rockland County Solid Waste Management Authority/ Sewer District

North Carolina
Charlotte/Mecklenburg Utilities
Durham, City of
Metropolitan Sewerage District of Buncombe County
Orange Water & Sewer Authority

Ohio
Akron, City of
Butler County Department of Environmental Services
Columbus, City of
Metropolitan Sewer District of Greater Cincinnati
Northeast Ohio Regional Sewer District
Summit, County of

Oklahoma
Tulsa, City of

Oregon
Clean Water Services
Eugene/Springfield Water Pollution Control
Water Environment Services

Pennsylvania
Philadelphia, City of
University Area Joint Authority, State College

South Carolina
Charleston Commissioners of Public Works
Mount Pleasant Waterworks & Sewer Commission
Spartanburg Sanitary Sewer District

Tennessee
Cleveland, City of
Knoxville Utilities Board
Murfreesboro Water & Sewer Department
Nashville Metro Water Services

Texas
Austin, City of
Dallas Water Utilities
Denton, City of
El Paso Water Utilities
Fort Worth, City of
Gulf Coast Waste Disposal Authority
Houston, City of
San Antonio Water System
Trinity River Authority

WERF SUBSCRIBERS

Utah
Salt Lake City Corporation

Virginia
Alexandria Sanitation Authority
Arlington, County of
Fairfax County Virginia
Hampton Roads Sanitation District
Henrico, County of
Hopewell Regional Wastewater Treatment Facility
Loudoun County Sanitation Authority
Lynchburg Regional WWTP
Prince William County Service Authority
Richmond, City of
Rivanna Water & Sewer Authority

Washington
Edmonds, City of
Everett, City of
King County Department of Natural Resources
Seattle Public Utilities
Sunnyside, Port of
Yakima, City of

Wisconsin
Green Bay Metro Sewerage District
Madison Metropolitan Sewerage District
Milwaukee Metropolitan Sewerage District
Racine, City of
Sheboygan Regional Wastewater Treatment
Wausau Water Works

Australia
South Australian Water Corp.
Sydney Water Corp.
Water Corp. of Western Australia

Canada
Greater Vancouver Regional District
Toronto, City of, Ontario
Winnipeg, City of, Manitoba

Mexico
Servicios de Agua y Drenaje de Monterrey, I.P.D.

New Zealand
Watercare Services Limited

Singapore
Singapore Public Utilities Board

United Kingdom
Yorkshire Water Services Limited

STORMWATER UTILITY

California
Los Angeles, City of, Department of Public Works
Monterey, City of
San Francisco, City & County of
Santa Rosa, City of
Sunnyvale, City of

Colorado
Boulder, City of

Georgia
Griffin, City of

Iowa
Cedar Rapids Wastewater Facility
Des Moines Metro Wastewater Reclamation Authority

Kansas
Overland Park, City of

Kentucky
Louisville & Jefferson County Metropolitan Sewer District

Maine
Portland Water District

Minnesota
Western Lake Superior Sanitary District

North Carolina
Charlotte, City of, Stormwater Services

Pennsylvania
Philadelphia, City of

Tennessee
Chattanooga Storm Water Management

Washington
Seattle Public Utilities

STATE

Arkansas Department of Environmental Quality
Fresno Metropolitan Flood Control District, Calif.
Urban Drainage & Flood Control District, Colo.

CORPORATE

ADS Environmental Services
The ADVENT Group Inc.
Alan Plummer & Associates
Alden Research Laboratory
Aqua-Aerobic Systems, Inc.
Aquateam–Norwegian Water Technology Centre A/S
BaySaver Inc.
BioVir Laboratories, Inc.
Black & Veatch
Boyle Engineering Corporation
Brown & Caldwell
Burns & McDonnell
CABE Associates Inc.
The Cadmus Group
Camp Dresser & McKee Inc.
Carollo Engineers Inc.
Carpenter Environmental Associates Inc.
CDS Technologies Inc.
Chemtrac Systems Inc.
CH2M HILL
Damon S. Williams Associates, LLC
David L. Sheridan, P.C.
Dewling Associates, Inc.
Earth Tech Inc.
Eco-Matrix
Ecovation
Environmental Engineers International
EMA Inc.
The Eshelman Company Inc.
Finkbeiner, Pettis, & Strout Inc.
Freese & Nichols, Inc.
ftn Associates Inc.
Fuss & O'Neill, Inc.
Gannett Fleming Inc.
Golder Associates Inc.
Greeley and Hansen LLC
Hazen & Sawyer, P.C.
HDR Engineering Inc.
HNTB Corporation
HydroQual Inc.
Infilco Degremont Inc.
Ingersoll-Rand Energy Systems
Insituform Technologies Inc.
Institute for Environmental Technology & Industry, Korea
Jacobson Helgoth Consultants Inc.
Jason Consultants Inc.
Jordan, Jones, & Goulding Inc.
KCI Technologies Inc.
Kelly & Weaver, P.C.
Kennedy/Jenks Consultants
Komline Sanderson Engineering Corporation
Lawler, Matusky & Skelly Engineers, LLP
Limno-Tech Inc.
Lombardo Associates Inc.
Malcolm Pirnie Inc.
McKim & Creed
MEC Analytical Systems Inc.
Metcalf & Eddy Inc.
MWH
New England Organics
Odor & Corrosion Technology Consultants Inc. (OCTC)
Oswald Green, LLC
PA Government Services Inc.
Parametrix Inc.
Parsons
Post, Buckley, Schuh & Jernigan
R&D Engineering/Conestoga Rover & Associates
The RETEC Group
RMC, Inc. (Raines, Melton & Carella)
R.M. Towill Corporation
Ross & Associates Ltd.
Royce Technologies
SAIC Maritime Technical Group
Short Elliott Hendrickson, Inc.
Stantec Consulting Group, Inc.
Stormwater Management, Inc.
Synagro Technologies, Inc.
Tetra Tech Inc.
Trojan Technologies Inc.
URS Corporation
Veolia Water NATC
Wade-Trim Inc.
Weston Solutions Inc.
Woodard & Curran
WRc/D&B, LLC
WWETCO, LLC
Zoeller Pump Company

INDUSTRY

American Electric Power
ChevronTexaco Energy Research & Technology Company
The Coca-Cola Company
Dow Chemical Company
DuPont Company
Eastman Kodak Company
Eli Lilly & Company
Merck & Company Inc.
ONDEO Services
Procter & Gamble Company
PSEG Services Corp.
RWE Thames Water Plc
Severn Trent Services Inc.
United Water Services LLC

Note: List as of 7/1/04

Board of Directors

Chair
James F. Stahl
County Sanitation Districts of Los Angeles County

Vice-Chair
Vernon D. Lucy
ONDEO Degremont Inc.

Secretary
William J. Bertera
Water Environment Federation

Treasurer
Karl W. Mueldener
Kansas Department of Health & Environment

Mary E. Buzby, Ph.D.
Merck & Company Inc.

Dennis M. Diemer, P.E.
East Bay Municipal Utility District

Jerry N. Johnson
District of Columbia Water and Sewer Authority

Richard D. Kuchenrither, Ph.D.
Black & Veatch

Alfonso R. Lopez
New York City Department of Environmental Protection

Richard G. Luthy, Ph.D.
Stanford University

John T. Novak, Ph.D.
Virginia Polytechnic Institute & State University

Lynn H. Orphan
Kennedy/Jenks Consultants

J. Michael Read
HDR, Inc.

James M. Tarpy
Nashville Metro Water Services

Murli Tolaney
MWH

Executive Director
Glenn Reinhardt

Research Council

Chair
John Thomas Novak, Ph.D.
Virginia Polytechnic Institute & State University

Vice-Chair
Glen T. Daigger, Ph.D.
CH2M HILL

Robert G. Arnold, Ph.D.
University of Arizona, Tucson

Robin L. Autenrieth, Ph.D.
Texas A&M University

William L. Cairns, Ph.D.
Trojan Technologies, Inc.

James Crook, Ph.D.
Water Reuse Consultant

Geoffrey H. Grubbs
U.S. EPA

Mary A. Lappin, P.E.
Kansas City Water Services Department

Keith J. Linn
Northeast Ohio Regional Sewer District

Drew C. McAvoy, Ph.D.
Procter & Gamble Company

Margaret H. Nellor, P.E.
County Sanitation Districts of Los Angeles County

Spyros Pavlostathis, Ph.D.
Georgia Institute of Technology

Steven M. Rogowski, P.E.
Metro Wastewater Reclamation District of Denver

Peter J. Ruffier
Eugene/Springfield Water Pollution Control

Michael W. Sweeney, Ph.D.
EMA Inc.

George Tchobanoglous, Ph.D.
Tchobanoglous Consulting

Gary Toranzos, Ph.D.
University of Puerto Rico

Stormwater Technical Advisory Committee

Chair
Robert E. Pitt, Ph.D., P.E., D.E.E.
University of Alabama

Vice-Chair
Ben Urbonas, P.E.
Urban Drainage and Flood Control District

Christine Andersen, P.E.
City of Long Beach, California

Gail B. Boyd
URS Corporation

Larry Coffman
Prince George's County

Brian Marengo, P.E.
City of Philadelphia Water Department

A. Charles Rowney, Ph.D.
Camp Dresser & McKee Inc.

James Wheeler, P.E.
U.S. EPA

WERF Product Order Form

As a benefit of joining the Water Environment Research Foundation, subscribers are entitled to receive one complimentary copy of all final reports and other products. Additional copies are available at cost (usually $10). To order your complimentary copy of a report, please write "free" in the unit price column. WERF keeps track of all orders. If the charge differs from what is shown here, we will call to confirm the total before processing.

Name ______________________ Title ______________________

Organization ______________________

Address ______________________

City ________ State ________ Zip Code ________ Country ________

Phone ________ Fax ________ Email ________

Stock #	Product	Quantity	Unit Price	Total
			Postage & Handling	
			VA Residents Add 4.5% Sales Tax	
			Canadian Residents Add 7% GST	
			TOTAL	

Method of Payment: (All orders must be prepaid.)

❑ Check or Money Order Enclosed

❑ Visa ❑ Mastercard ❑ American Express

Account No. ________ Exp. Date ________

Signature ________

Shipping & Handling:

Amount of Order	United States	Canada & Mexico	All Others
Up to but not more than:	Add:	Add:	Add:
$20.00	$5.00*	$8.00	50% of amount
30.00	5.50	8.00	40% of amount
40.00	6.00	8.00	
50.00	6.50	14.00	
60.00	7.00	14.00	
80.00	8.00	14.00	
100.00	10.00	21.00	
150.00	12.50	28.00	
200.00	15.00	35.00	
More than $200.00	Add 20% of order	Add 20% of order	

*minimum amount for all orders

Note: Please make checks payable to the Water Environment Research Foundation.

To Order (Subscribers Only):

Log on to www.werf.org and click on the "Product Catalog."

Phone: (703) 684-2470
Fax: (703) 299-0742.

WERF
Attn: Subscriber Services
635 Slaters Lane
Alexandria, VA 22314-1177

To Order (Non-Subscribers):

Non-subscribers may be able to order WERF publications either through WEF (www.wef.org) or IWAP (www.iwapublishing.com). Visit WERF's website at www.werf.org for details.

IWA Publishing's authorised EU representative for General Product Safety Regulations is Diane D'Arras, 15 rue Duret, 75116 Paris, France, e-mail: safety@iwap.co.uk.

Printed and bound by CPI Group (UK) Ltd, Croydon, CR0 4YY
12/05/2026
02108879-0008